Hill Country Backroads

Showing the Way in Comal County

Joe Sanders. 1894–1964.

To My Family

Hill Country Backroads

Showing the Way in Comal County

Laurie E. Jasinski

with an afterword by

Laurie P. Sanders

TEXAS CHRISTIAN UNIVERSITY PRESS / FORT WORTH

Library of Congress Cataloging-in-Publication Data

Jasinski, Laurie E.

Hill country backroads : showing the way in Comal County / Laurie E. Jasinski ; with an afterword by Laurie P. Sanders.

p.cm.

Includes bibliographical references and index.

ISBN 0-87565-239-5 (alk. paper)

1. Roads—Texas—Comal Co.—History. 2. Automobile travel—Texas—Comal Co.—History. I. Title.

HE356.T4 J37 2001.

388.1'2—dc21

00-049443

Cover and text design by Bill Maize—Duo Design Group.

CONTENTS

PREFACE

Hill Country Backroads has truly represented a journey for me from the very first seeds of thought straight through to the book's completion. I never knew my grandfather Joe Sanders. He died three months before I was born and so was always a ghost of others' memories—a symbolic figure. I could only picture from stories that he was a hardworking, genuine man who devoted himself to community service. He also enjoyed the simple pleasures of a country drive—just like my family, years later, when some summer evening Daddy would suggest, "Let's go for a ride," and I at age five or six or seven would watch the world transform beyond the car window into an endless countryside of hills and tree lines and green pastures with cattle or deer. Once in a great while I heard mention of Joe's adventures down the county backroads.

Over a decade ago, I asked my grandmother Laurie P. Sanders about Joe's accomplishments. She went digging through drawers and closets and presented scrapbooks, letters, maps, and albums of memories. She was eighty-four then, and I'd wished that I'd asked much sooner. She brought forth a wealth of information that revealed to me not only pieces of Joe's life, but a view into the lost world of early travel in the Texas Hill Country.

In the early twentieth century, the quick comfortable trip down the paved divided highway did not exist. The automobile was a novelty, and roads, especially the scenic country byways, were little more than trails—rough, wild, and unmarked. Taking that Sunday drive was surely an adventure, but as the car progressed to become both the great workhorse and toy of the twentieth century, motorists called for the improvement of roads. The age of the auto tourist was beginning.

The Texas Hill Country of Comal County and beyond was beautiful but largely inaccessible to the traveler. With no road maps or signs, drivers were at the mercy of their wits or the directions from some kind local they might happen upon. This prompted Joe Sanders to take action by posting road signs and compiling the first scenic road map of Comal County.

My grandmother was with him every step of the way, and it is in her words that the joys and struggles of those drives in the 1920s and 1930s come to life. It is in the words of many others as well—individuals who shared their stories with me—and as their words came alive the past became real to me.

One sunny Friday in April 1988, I piled Joe's maps, a notebook and my dog into my car and drove many of the same roads he had decades before. They were all paved, well-maintained, and, compared to his map mileage figures, the modern distances were consistently shorter. Today we live in a more direct, fast-paced world. Many of the extra curves and twists of the leisurely old backroad are gone. Amazing how fast things change.

I have tried to capture some of the essence of early travel and tourism in this book. Perhaps for some readers it will rekindle fond recollections of driving along the river in Dad's old Model T or struggling to power up a hill before the old car boiled over. For younger readers, raised in the age of the superhighway, I hope that this story will open a window to the past and highlight how we take for granted today's convenient travel. This concept had never even occurred to me until I started my research and talked to people who journeyed down the old byways.

While writing this book I wanted to tell an interesting story of the early backroads and scenic destinations, of drivers' struggles, and my grandfather's role in helping to solve them. Though much information relates to local events, traveling troubles occurred nationwide, so I included facts that also present a broader perspective, making Comal County a microcosm for the changing times in general.

I am very fortunate that so many of Joe's photographs and documents were preserved. This book contains many of them, including three of his county maps. Though he issued a series of editions from 1933 to 1960, I chose to include those maps that I feel are the most appealing and representative of his compilations. The 1933 map was the original edition. The 1936 Centennial Edition is my favorite and was apparently a favorite of many tourists. In my opinion it is probably most illustrative of the time, with its marked scenic tourist attractions, sketches, and colorful comments. The 1960 map was the last map and very evident of the sweeping changes taking place in Comal County.

Throughout all of this work with its joys and delays and fascinating discoveries, I have been blessed with a loving and supportive family. My parents Richard and Laurie Jasinski have exemplified the power of belief and perseverance and have offered helpful advice. Thanks to my brother Larry Jasinski, who actually knew Joe Sanders as his grandfather, for his stories and for accompanying me on the occasional field trip. I am grateful to my husband Gary Hickinbotham for his valuable observations, companionship on all those scenic excursions, and for planting the seed to write a book in the first place. Most important my deepest thanks to my grandmother Laurie P. Sanders for her memories, all those wonderful interviews, and the endless but fun hashing out of details. This book would not be possible without her. And, of course, thanks to my grandfather for his dreams and accomplishments.

In many ways Joe Sanders really did show the way through his signs and maps, but most importantly through his service. It has been a remarkable experience—to interview fascinating characters, research forgotten facts, and track down the scenic hideaways. In so doing, here I am years later, in some of Joe Sanders's same old haunts. But my grandfather is no longer a ghost. He showed me the way.

ACKNOWLEDGMENTS

Many people provided valuable information and insightful anecdotes through personal interviews, conversations, and "Reflections" radio program transcripts, as cited in the bibliography. My deepest appreciation to everyone, especially those who furnished ample quotes and facts: Milton Kaderli, Larry Kearney, Dorothy Wimberley Kerbow, William Kolodzie, the Alfred Liebscher family, Hip Mengden, Melba Meckel Roth ("Reflections" program), and Rudy Seidel. A very special thanks to Donald W. Olson, professor of physics at Southwest Texas State University, for his careful reading and editing of draft after draft of this book, and to Kelly A. S. Borsheim for her photography services including shooting copy negatives, printing, and photo retouching.

Thanks to the following individuals, research institutions, museums, and scenic parks: Alamo Classic Car Showcase and Museum, New Braunfels, Texas; Albert B. Alkek Library, Southwest Texas State University, San Marcos, Texas; American Concrete Institute, Detroit, Michigan; American Legion, Comal Post 179, New Braunfels, Texas; Blanco County Clerk's Office, Johnson City, Texas; Blanco Library, Blanco, Texas; Charlotte Boyd; Cascade Caverns, Boerne, Texas; Center for American History (formerly Barker Texas History Center), University of Texas at Austin; Comal County Clerk's Office, New Braunfels, Texas; Comal County Commissioners Court, New Braunfels, Texas; Comal County Road Department, New Braunfels, Texas; Dr. Christopher S. Davies of the Geography Department, University of Texas at Austin; Eugene Ebell at Cave Without A Name, Boerne, Texas; Pépé Gutierrez at the Seven Eleven Ranches; Mrs. Hal Harwell; Mark Havers at Slumber Falls Camp; Hays County Clerk's Office, San Marcos, Texas; Evelyn Heckel; Clyde Jacobs; David Jonas; Kendalia Library, Kendalia, Texas; Lawrence Krause; Wesley Meyer; Natural Bridge

Caverns, Comal County, Texas; Portland Cement Association, Skokie, Illinois; Ann Rogers; San Marcos Public Library, San Marcos, Texas; Iris Schumann; Sophienburg Archives, New Braunfels, Texas; Texas Department of Transportation, Austin, Texas; Texas State Historical Association, Austin, Texas; Mrs. J. C. Wessendorff.

. . . . Part 1 . . . .

THE EARLY DAYS OF MOTORING AND DREAMS

"... the journey,

with all

the struggles

involved ..."

Chapter One

Take to the Highway

In 1909 a strange, noisy, smoking machine puttered down a dirt road that sliced through golden Ohio grainfields. The motorized horseless buggy amazed onlookers and left billowing clouds of dust and fumes in its wake. As the driver pulled up to a farmhouse a wide-eyed fourteen-year-old boy greeted him excitedly. Young Joe Sanders lived with his grandparents on their farm near the little town of Ottoville in northwestern Ohio. The agricultural landscape consisted of flat expanse, with cornfields stretching to the horizon, and a boy's life meant hard work, schooling, and church on Sunday. There was little time for frills or diversions. Grandma's "waste not, want not" and "haste makes waste" ruled the day. So it caused quite a stir when a family friend proudly arrived in a motorcar instead of a wagon.

For the first time Joe Sanders saw an auto, a 1904 Oldsmobile, and the sputtering contraption fascinated him. He asked if he could drive–just around the farm. First the man had to explain the controls, but in no time Joe was off. The car bounced down the country roads and stopped wagons in their tracks as the horses reared up in panic. All along the way Joe caught the attention of neighboring country folk, and he could not resist the urge to drive into town and show off his newly discovered ability to pilot the strange machine. By the time he drove back home he was determined to learn everything about motorcars. His mechanical mind was unleashed. That day he was hooked, and the automobile for him would forever be a means of enjoyment, innovation, and growth.[1]

✦ ✦ ✦ ✦

In the early twentieth century automobiles provided new and exciting opportunities for business, travel, and recreation. Many towns catered to a new concept–tourism. With a car, in theory, the average family could tour the countryside, visit friends in another town, stop for a picnic, and be home by evening. In practice, early motoring had many pitfalls. Autos suffered mechanical failures, and poor road conditions hindered transportation.

Across the United States the development of travel was a slow process of trial and error as the needs of travelers changed. In the vast area of Texas, journeying from town to town could involve great distances over varying country. The earliest roads followed Indian trails cutting through the humid pine forests of East Texas, the flat Coastal Plains, the grassland prairies to the north, and the mountains of the arid west. Some of these passages traced the route of buffalo herds following the course of least resistance around rough terrain. In the Central Texas Hill Country, much of the landscape was composed of limestone. Its porous consistency drained fairly well, but the rock provided a rugged path that could be difficult to traverse. Wagon drivers often had to detour around hills rather than struggle over them, and wayfarers carefully forded the shallowest, most secure river crossings.

In the late 1600s and during the 1700s, Spanish explorers carved a network of roadways across the Texas wilderness, further defining and linking the existing trails of the native inhabitants. El Camino Real, an early road blazed by several expeditions in the 1690s, ran from the Rio Grande southeast of present-day Eagle Pass, through Nacogdoches, and east toward the Sabine River. The course, commonly known as the Old San Antonio Road, served as a major artery through Texas and was the only connecting route for many towns and missions. Anglo-American pioneers who moved into Texas beginning in the 1820s made extensive use of the early roads.[2]

In 1845 German settlers landed at the Texas Gulf Coast and trekked inland to the Hill Country area that would become Comal County. They discovered a land of abundant wildlife and two rivers. The clear artesian Comal Springs welled out of the base of a limestone cliff and formed the

Comal River, which flowed three miles downstream into the Guadalupe River. Here the pioneers founded the city of New Braunfels, and, realizing the need for adequate roads, on August 7, 1846, they organized the Comal County Commissioners Court.

The court consisted of four commissioners, one for each precinct in the county. They had the power to establish and lay out new courses, discontinue old routes, seize property, and build roads. The commissioners wasted no time in providing for an orderly network of avenues for citizens. New thoroughfares extended from the lonely segment of the Old San Antonio Road (part of present Nacogdoches Street) on the edge of town. As early as January 12, 1847, the court declared the Seguin and San Antonio streets in New Braunfels to be highway roads. In June 1847, the commissioners agreed to lay out a route to Fredericksburg, and work started later that year. Interestingly, the court received a bill on February 7, 1848, requesting payment for a packhorse. The road overseer required a horse for five days for viewing the road toward Fredericksburg. He charged 75 cents a day.[3]

Planning and laying out a roadway was a complex endeavor. Citizens had to petition the commissioners court to open a new route or make changes on an existing one. The commissioners then appointed a jury of view that determined the feasibility and necessity of the petitioners' request. Occasionally the jury saw no need for a road, but most of the time the court carried out the plans. When a road went through, landowners were compensated for property and damages. The going rate for land compensation by 1889 was roughly $10 per one-half acre and $25 for one acre.

Sometimes owners objected to a new road and refused to yield their land. The commissioners court often went "as a body" to view an area in order to settle disputes about right-of-way. Property owners also charged the court for extra damages they felt they deserved. In the late 1880s, Henry Timmermann protested that the new road going past the International and Great Northern railroad bridge outside of town ran too close to his family's burial ground. The court awarded him $355, but the road work continued.

In 1887 commissioners ordered rancher Henry Pantermuehl to remove a gate obstructing Mountain Valley Road, and it was the duty of road overseer George Fischer to keep the byway clear. They ordered Henry Dietz to make openings in his rock fence along the New Braunfels-Fredericksburg Road to improve water drainage.[4]

Water lanes played an important role in road development in that they allowed landowners access to the nearest spring, creek, or other body of water. A blocked water lane could be the cause of a heated argument. Any establishment of or change in water routes had to get the approval of the commissioners court just like other roads.[5]

By the latter part of the 1800s some established routes were already in use in Comal County including the Smithson's Valley-Boerne Road, Bear Creek Road, Blanco-San Antonio Road, Cranes Mill Road, Purgatory Road, and Mountain Valley Road. Usually the road names reflected nearby destinations, but many water crossings were named after the people who lived by them. The best identifiable markers along byways before the age of signs might be "So 'n So's" gate or barn.

Roads required constant maintenance that consisted of long hours chopping weeds and branches, clearing away large stones, digging drainage ditches and tunnels, spreading gravel, and performing other strenuous tasks. Rough terrain severely limited the work. In June 1887 authorities changed part of the course of Smithson's Valley-Boerne Road to "avoid a rocky hill and a boggy place on the old road."[6] Sometimes steep ledges and deep ravines caused workers to stop short altogether.

The commissioners court appointed overseers to supervise maintenance. These foremen were free to name whom they saw fit, usually residents along the stretch, to work on the backroads and farm trails. The property owner could work or pay someone else to do it for him. Up into the early 1900s the court required "all able bodied male persons between the ages of 21 and 45 years" to perform road duties. They had to volunteer at least several days a year working on a road within the precinct where they lived. The court later even declared that all people convicted of misdemeanors and punished by fine who failed to pay would be "compelled to work out the amount of such judgment on the public roads of Comal County...."[7]

The county paid for work on the main public thoroughfares and appropriated $25 in August 1889 for repairs on the San Antonio-Blanco Road. Wages certainly reflected the times. By 1900 hands working on public roads were paid $1.25 per day. Drivers of a two-horse team could earn $2.50 per day. A four-horse team earned $3.50.[8]

At the turn of the century, the business of local highway development was taking place in Comal County and other counties across Texas. Two million miles of roads zigzagged the United States then, but of that number only 150 miles had some kind of hard pavement, and that was in large cities.[9] Most roads were nothing more than pitted rocky trails or soggy mudholes, and with this new invention, the automobile, taking the country by storm, public officials took notice.

In 1895 just four automobiles were registered in the United States. Henry Ford operated his first car on June 4, 1896. Later that year pioneer racers held the first United States auto race on a closed course. An electric vehicle, the Riker Electric Stanhope, won the event. However, the chant "Get a Horse!" would remain a very popular saying for some years; and drivers often needed a team of horses to pull a vehicle out of the mud. Even so, in 1897 the number of cars registered had grown to ninety nationwide, and by 1899 the figure was 3,000. Motorists increasingly lobbied to make road building a top priority. In Texas, road enthusiasts considered establishing a Bureau of Highways as early as 1903, but nothing was carried out.[10] The responsibility still rested at the county level where available money came out of a road-and-bridge fund. In 1903 the court in Comal County issued $25,000 in bonds, and in 1904 officials debated over whether to levy a 5-cent road-and-bridge tax on every $100 valuation of property in the county.[11]

On April 10, 1904, a *Dallas Morning News* correspondent commended the streets of New Braunfels in a special feature titled "Life in Quaint Old German Town, Whose People Are Ignorant of Poverty and Destitution":

> They are kept as clean as the floors in the home of a good housewife. Broad walks on either side are covered with gravel or flagstones, hedged in on the one hand by fences over which flowers peep and on the other by trees in which the mocking birds nest and sing.... There are men who look after the streets. By 8 o'clock each morning New Braunfels is as clean as a pin and that way it remains....

The story heralded the quaint German town as a prosperous old world place where "rosy-cheeked girls and portly old women" tended their yards with "jealous care" while the men were "sweeping out the rubbish from the business houses, cleaning the sidewalks and burning trash in the gutters." The feature stated, "...there is apparent [sic] no good reason why it should not become more popular with tourists."

New Braunfels was already gaining a reputation as a winter resort with a mild climate. Landa Park, located along the crystal headwaters of the Comal River, was renowned as one of the most beautiful attractions in the Southwest, and every weekend tourist trains brought picnickers. The Missouri, Kansas, and Texas railroad advertised round-trip tickets in the *San Antonio Daily Express* on May 1, 1904, for 50 and 25 cents.

In 1907 Landa Park owner Harry Landa became one of the earliest New Braunfels citizens to buy a car. His county license plate number was "4." In those days auto drivers had to yield the right-of-way to horses. The vehicles often spooked the animals. In his book, *As I Remember...*, Harry Landa commented that the novelty of the motorcar alarmed farmers. Consequently, many rural residents of Comal County petitioned the commissioners court to "pass a resolution ruling me off the road," Landa wrote, "because they said the gas machine would frighten their teams when they met on the road."[12]

In the early 1900s farmers still drove their wagons into town on Saturdays to stock up on supplies. As a boy, Comal County native Milton Kaderli recalled his father's stories of seeing the weekly pilgrimage from the Hill Country. The farmers would spend the night in New Braunfels, camping in one of the merchant's yards–behind the Louie Henne

The lush greenery and serene waters of Landa Park and the Comal River area invited tourists and townspeople alike to partake in a relaxing picnic or boat ride. (Photo by Joe Sanders. Unless otherwise indicated copies of all illustrations are in the possession of Laurie E. Jasinski.)

Company or the Pfeuffer Store, for example. Kaderli said, "They had a backyard there where customers could park, leave their vehicles, feed their animals, and sleep there during the night." After conducting their business, they headed home the next day.[13]

As a general rule, a farmer might make a trip once a month. Kaderli added, "It was a long ways and very inconvenient, and they would put off a trip to town until they just absolutely had to go."

An excerpt from an article about the Pfeuffer Store published in *The Inland Merchant* traced the history of wagons coming to town:

> Back in the years before the advent of the automobile and good roads, commodious hitching space for customers who remained in town only during the day, and adequate sheds for those who remained over night, was vitally important. Accordingly, this company established a large space directly back of the store, with hitching posts along the sides; and across one end they erected water-proof sheds for wagons and teams

> whose owners found it necessary to remain in town during the night. These sheds not only protected the wagons and teams, but provided a camping place for the owners of them as well. The place was an all round "wagon yard" provided by the store at no charge.[14]

Good Roads Associations formed in Texas and across the country as early as 1903 and played a key role throughout the early twentieth century in promoting better highways. By 1906 the United States government finally opened its Office of Public Roads with a budget of $37,660. Some Texans wanted to follow suit in 1911 and establish a state highway department by levying a license fee on autos, but the Texas public was not ready for it, so highway associations continued to push for action by organizing events like good roads days and auto tours.[15] On July 25, 1912, the *San Antonio Express* reported one such event—a tour arranged for farmers and ranchers dubbed "The Farm and Ranch Tour."

A total of twenty-six vehicles motored out of Dallas on July 22. By the afternoon of July 24, twenty-three arrived from Austin and drove through the business section of San Antonio. The *Express* explained:

> Primarily, this tour is to increase the interest in good roads throughout the State and one of the conditions of the long run is that owners of the cars must drive all the way and are not permitted to have any substitutes. All of the contesting cars carried passengers. If it was a runabout there were two or more persons on board and a touring car called for four or more persons. The dust-covered cars, some of them ghostly in appearance, were parked last night on Alamo Plaza and at 9 o'clock this morning their noses will again be headed northward to reel off the miles back to the home base....
>
> The tour is voted a big success....[16]

Citizens became increasingly aware of the need for better roads even though there were still only some 250 miles of concrete highways nationwide in 1912. Travelers pushed to improve the routes between towns, and the Comal County Commissioners Court paid laborers working on public thoroughfares $1.50 for an eight-hour day; a two-horse team and driver earned $3.00.[17]

On October 8, 1913, the *San Antonio Express* publicized a Good Roads Day "set aside for building the San Antonio-Austin highway." The San Antonio area was especially active in its support of road improvement, and citizen David E. Colp was a full-time promoter of the Texas Good Roads Association. He assisted in the formation of more than 100 local roads clubs. As secretary of the Bexar County Highway League, Colp was instrumental in organizing the Good Roads Day, and he lobbied civic groups from San Antonio to Austin to support the event. Scheduled for November 5, 1913, the goal was to enlist thousands of volunteers to work along the entire stretch of road, and committees visited residents living within two or three miles of the route. The *Express* commented, "Not a man will be overlooked. He will be requested either to put on his overalls, take his shovel and help, or hire someone to take his place."

Martin Faust, secretary of the Good Roads Club at New Braunfels, was in charge of securing help in his area by traveling "overland along the route within a short time preaching to the farmers the gospel of better roads and arousing interest in the project."

Like other towns across Texas, the motorcar transformed New Braunfels. Around the 1910s, crews improved city streets by a process of graveling called macadamizing. Workers packed the roads with layers of progressively smaller rocks until the top layer consisted of crushed stones called screenings no larger than two inches in diameter, but an unpleasant result of this was dust—lots of it. Consequently a watering cart sprinkled down the road everyday.[18]

The emergence of the car also changed the face of downtown businesses such as the Pfeuffer Store, which gradually removed some hitching posts and made space for automobiles. "The combination of good roads and cheaper automobiles..." the *Inland Merchant* reported, "has practically transformed the old-time 'wagon yard' of the early days into a modern automobile parking lot today." A few sheds and hitching posts remained, however, to accommodate customers who used teams and wagons.

The first auto dealership in town was Gerlich Auto Company, which opened in 1912 and sold Model T Fords. By 1913 Ford sales ads boasted that two million families needed a motorcar for pleasure, utility, and family activities. The early auto companies and garages provided a wide range of services. Car salesmen often took in animals, wagons, and even food for trade. They sold gasoline, kerosene, and carbide as fuel for the engines and headlights of that time. In the early days, car lights operated with gas or carbide generators or sometimes by kerosene oil lamps. Therefore, it was vital for motorists to keep an adequate supply of lighting fuel on hand to avoid a catastrophe while driving at night.[19]

A long-time New Braunfels auto dealer, Johnny Ruppel, arrived in the small German town as a teenager in 1913 and went into the car business. He remembered the dedication required to work in a garage. "It developed into a 24-hour job because I had to have a room at the garage so that I could take care of the night work." Customers could ring a bell, summoning the young man at all hours of the night. Many of his late-night calls involved desperate drivers needing fuel to replenish their fading lights. After his employment at the garage, he worked at New Braunfels Auto Sales and Service, another dealership that set up shop, selling Maxwell cars and eventually Chevrolet, Monroe, and Buick models. Ruppel recalled that they "possibly sold 14 or 16 automobiles in a year." As for servicing the vehicles, he added, "We were always happy when they started after we'd worked on them."[20]

On May 8, 1914, the *New Braunfels Herald* printed a front-page story about the "model federal highway to be built from Austin to San Antonio." The "Federal Post Road," as it was then called, was touted to be the first of its kind in Texas. The stretch from Austin to San Antonio, still little more than a "wagon road," would be replaced with a roadway that was sixteen feet wide with a layer of twelve inches of gravel. Financing for the new eighty-one-mile stretch was to come from a combination of funds from Travis, Hays, Comal, and Bexar counties, totaling $140,000, and available money from the United States Department of Agriculture, which would supervise the project. Within a month Comal County had proudly set $44,750 aside for its share of construction costs, making it the first county to have the necessary funds appropriated. The total figure included $39,000 in county money from road-and-bridge bonds and $5,750 in donations from prominent citizens George Eiband, S. V. Pfeuffer, and Joseph Faust. The county already had previously acquired significant road-building equipment such as a ten-ton gasoline road roller purchased for $3,000 in 1911, a stone-crushing outfit for $2,025 and a road harrow known as a "scarifier" for $85 in 1914.[21]

On October 20, 1914, a groundbreaking ceremony for the new Austin-San Antonio Post Road (early forerunner to Interstate 35) took place at the hamlet of Hunter in Comal County. The affair represented a triumphant victory for good roads advocates and was an inter-county celebration with a brass band, barbecue dinner, and numerous speakers. Comal County attorney Martin Faust, who had worked so diligently a year earlier to generate awareness and enthusiasm for better roads, was among the speakers.

Governor Oscar Colquitt "plowed the first furrow for the great highway," and the *New Braunfels Herald* reported on the stirring sentiment of the scene of the state governor carving through the first 100 yards of earth:

> ...he must have felt that steadying a big road plow drawn by six restless mules, and a hard and sticky black soil is no less difficult a job as guiding the ship of state, drawn as it were, by one hundred seventy odd obstinate legislators.... The sight of this will remain a pleasant remembrance to all present; and it makes a historical event for the town of Hunter.[22]

The awarded bid for construction of Comal County's portion of the Post Road was $60,000 to road builder A. M. Clopton. A United States senior highway engineer oversaw the project. Work on the Austin-San Antonio

Post Road was complete in March 1916.[25] This artery would help open the area to more wayfarers. New Braunfels was poised on the edge of discovery by the motoring tourist.

Joe Sanders at age eighteen. (Laurie P. Sanders Collection.)

In late 1916 the young man from Ohio arrived in New Braunfels. He was weakened from a recent illness and just a bit homesick for his grandparents' farm during the Christmas season. But he was also curious and ready for new discoveries, and the neat little German community seemed like an interesting place with its orderly streets, surrounding wild countryside, and excursion trains into the park.

Joe Sanders had spent the previous few years trying to make it in the world, honing his natural talents along the way. At the age of sixteen he had left the Ohio farm and traveled forty miles west to Fort Wayne, Indiana, to find work. For four years he labored at a variety of jobs including employment with the Fort Wayne Corrugated Paper Company, General Electric, the Wabash and Pennsylvania railroads, and the Perfection Biscuit Company. Then he moved to Toledo, Ohio, around 1915, and sharpened his mechanical skills at the Overland Auto Factory and the Consolidated Motor Company.

Joe was a conscientious hard-working perfectionist who heeded the strong work ethic of his grandparents. But the cold climate took its toll on the slender-built fellow, and in October 1916 he fell victim to typhoid fever. Pneumonia set in and a doctor urged Joe to move to a warmer climate for a while to recover.

The prospect of going to a new and different place with a mild climate appealed to Joe, and so friends around Ottoville set out to help him in his journey. Their local pastor had a relative, a fellow priest, who lived in New Braunfels, Texas. Surely he could help to get Joe established in a new place. A friend who was interested in a change of scene offered to accompany Joe. The two headed for Texas, arriving in New Braunfels on December 19, 1916.[24]

Chapter Two

New Cars, Open Roads, and a Young Adventurer

Like many young men of his time, Joe Sanders thought that "to go to Texas was the most wonderful thing in the world, because Texas had the reputation of being wild and woolly." He and his fellow Ohioan were made welcome by the Hippolyt (or, simply H.) Dittlingers, who were friends with the priest. The Dittlingers, a notable family in New Braunfels, had visitors quarters and were happy to accommodate the two travelers in a guest house. The young men soon went to work at an automobile shop in San Antonio. When the United States entered World War I, the two fellows stopped work one day and said, "Let's quit and join the army." They dropped their tools and left to enlist. Joe, however, failed his physical because of some remaining scars on his lungs from the pneumonia, but his friend passed and went on. Having already quit his job, Joe did not know what else to do but go back to New Braunfels. He saw Mr. Dittlinger again and secured a job as a mechanic at his flour mill.[1]

H. Dittlinger owned not only the flour mill but a lime kiln and rock quarry outside of town. At that time three quarries operated in Comal County—the Dittlinger, Altgelt, and Landa rock crushers. The county was fortunate to have a ready supply of natural materials that were put to use on the roads.[2]

The lime screenings used in road building were really the waste products left over from the quarrying process—the smallest pebbles left behind after the lime was sifted through screens. Road crews packed down layers of screenings, which made for a fairly good topping. Workers also paved byways with caliche, a lime and stone mixture. Though caliche roads generally drained better than other surfaces, the substance's varying clay content affected its quality. The higher clay caliche expanded in wet weather and cracked in dry weather—not good conditions for a road.[5]

The number of automobiles increased dramatically during the first two decades of the twentieth century. While vehicle registrations had accounted for 3,000 cars nationwide in 1899, that number soared to 3,513,000 in 1916. That same year the United States government passed the Federal Aid Road Act, a measure ordained to create state highway departments and provide matching federal funds to build roads. The law inspired the Texas Legislature to establish the Texas Highway Department in 1917. Immediately the department opened a registration division and registered 194,720 vehicles statewide. Comal County was eager to apply for its share of funds. By December 1917 the commissioners court applied for state aid for the reconstruction of a stretch of the Austin-San Antonio Post Road—19.6 miles in Comal County that ran southwest to the Guadalupe County line. The Dittlinger Rock Crusher donated $5,000 worth of materials for the Post Road (also called Highway No. 2) in 1918, and the process of graveling required the scarification and removal of all stones bigger than two inches in diameter. By this time laborers received $1.75 for an eight-hour day, while a driver with a team of mules earned $3.50 a day.[4]

The Post Road was a major route for the delivery trucks of the Dittlinger flour mill, and Joe serviced and repaired them after long drives. He was happy to put his mechanical skills to work, but the Great War was still first and foremost in everyone's mind, and Joe, though grateful to have a decent job, decided after some months to try again to join the army. This time he passed his physical, and the army sent him to Austin where he entered the Camp Mabry Detachment of the University of Texas Automobile School. Afterward he was sent to Fort Omaha, Nebraska, where he served as a private in the 75th Balloon Company, a section of the U.S. Army Air Service. While there, he continued his education by studying correspondence courses. He also taught driving and mechanics to other soldiers.[5]

When the war was over, Joe returned to San Antonio to be discharged. He figured that he had put in his service. He had recovered his health and satisfied his curiosity about Texas. It was time to go home—back to Ohio.

In 1918, Private Joe Sanders was stationed at Fort Omaha, Nebraska, where he studied mechanics and taught the course to other army soldiers. (Laurie P. Sanders Collection.)

Joe Sanders—ready to assume chauffeuring duties for the Dittlinger family. (Laurie P. Sanders Collection.)

He decided to make one last visit to New Braunfels and bid goodbye to the Dittlinger family. But Dittlinger persuaded him to stay and offered him a job. By this time the flour mill had converted completely to truck service, whereas earlier some shipments had still come in by rail. Trusting Joe's mechanical expertise, Dittlinger wanted him to be in charge of keeping the fleet running. He also asked Joe to be a part-time chauffeur. Joe agreed, thinking that he would stay for a little while and eventually go home to Ohio (though he never did). He found a place to stay, and with a new job and a vehicle to drive, he was set to start again in New Braunfels.[6]

Joe became very familiar with the area. As H. Dittlinger's chauffeur, he was responsible for taking him on business trips to Austin and San Antonio by way of the Post Road. Reconstruction on the Comal County stretch was finished for now (being a part of approximately three million miles of U.S. roads with some kind of hard surface in 1919).[7] Interestingly, with the completion of work on Highway No. 2, many people took to exceeding the speed limit, which prompted the Comal County Commissioners Court in April 1919 to deputize a man with a motorcycle and employ him for $100 a month as a county traffic officer. The court also ordered the posting of two signboards that read: "Speed Limit 18 miles per hour, Violators will subject themselves to a fine of from $1.00 to $100.00."[8]

Even with improvements the Post Road was still a primitive highway. Roads in those days curved, zigzagged, took sudden right-angle turns and often went around hills or rocky creek beds. Highway construction did not always include smoothing the surface, and a vehicle could not go too fast because of all the dips. Some stretches of roads developed large ripples known as washboard. There were no barriers and railings to keep cars from skidding down into deep ditches along the side.[9] And a driver knew when he reached the Guadalupe County line because that portion of highway was totally unimproved. The *San Marcos Record* sarcastically referred to the stretch as having a "nation-wide reputation" of being "the worst piece of road on earth." This section was miles from the center of Guadalupe County, and that county had other more central thoroughfares to worry about. Dittlinger's son-in-law Alfred Liebscher often made the trip beginning in the 1920s. "They didn't pay any attention to those three miles [on Post Road]," he recalled. If there had been a heavy rain the crossing over Cibolo Creek was often flooded. A driver had to wait many long hours before the crossing was passable, and water stood there for days.[10] Liebscher recalled that a good trip to San Antonio took about an hour and a half, and it took about two and a half hours to Austin, where the highway was in poorer condition. "It just took time," he commented, "and we didn't know any better at that time, so it shows what you can live with if you don't know any better."

Post Road, of course, represented one of the best routes in Comal County. Many other roads, even first-class ones, were in worse condition. All roads at that time were classified as first, second, or third class, as established by the Fourteenth Texas Legislature in 1875. Early Texas law called for

As traffic increased, motorists realized the perils of driving, and the sharp turn or swollen crossing caused many a wreck. (Photo by Joe Sanders.)

the existence of first-class roads between county seats. Originally required to be "paths forty feet wide, cleared of timber,"[11] as major connections between towns they needed to be easily passable. This was not always the case. Post Road on the way to San Marcos got downright muddy and some sections had no topping. The highway to nearby Seguin to the east of New Braunfels became very rough past the Comal County line into Guadalupe County. Other routes were still under construction. In 1920 the Comal County Commissioners Court ordered a survey to complete construction of State Highway No. 46 northwest of New Braunfels. Before that time the route had been a series of stretches from one destination to another—the New Braunfels-Spring Branch Road, the Smithson's Valley-Blanco Road, for example. Eventually the court accepted a whopping bid of $79,802.85 for construction of 16.6 miles of the Smithson's Valley-Blanco Highway. The name of the road would be officially changed to the New Braunfels-Blanco Highway No. 46 (or simply Highway 46) and supported with funds left over from Highway No. 2. Work included stripping gravel pits and screening gravel, each at a rate of 15 cents a yard. Rolling the road ran $2.00 an hour, solid rock excavation was $1.75 a yard, and loose rock excavation cost thirty-five cents per yard.[12]

Second- and third-class byways were of lesser importance and given less priority. The county had only so much money, and road duty had been officially abandoned in February 1920. The county collected a tax of fifteen cents annually for road purposes. A second-class road was about thirty feet wide and usually led to a main highway. This classification included many city streets.[13]

All the other courses were third-class roads, only twenty feet in width and often almost totally unimproved. They were the country byways, meandering around every tree and filled with obstacles like branches and boulders. Bridges (if there were any) were constructed of large wooden beams that became slick with algae when the water went across them. It was on these backroads that Joe Sanders satisfied his sense of adventure, for they usually led to the prettiest scenery in the county—through the Guadalupe River Valley, to the little hamlets of Sattler, Bulverde or Fischer's Store, or to the ridge of hills known as the Devil's Backbone.

Being a chauffeur meant that Joe had access to cars, but he also had to service the Dittlinger family vehicles. They had three Franklins. The Franklin was one of the more expensive cars with a price starting at about $2,000. H. H. Franklin founded his company in 1902 in Syracuse, New York. The company called its product "The Car Beautiful"[14] and developed a small but faithful following consisting of professionals and engineers, especially aviators. (Charles Lindbergh and Amelia Earhart owned Franklins.) The Franklin was an air-cooled car with a wooden frame. The auto had no radiator and therefore had the advantage of a lighter weight and no radiator worries about water freezing in the cold and boiling over on hot days.[15] Air-cooled cars were rather unusual in town, and Dittlinger trusted Joe to maintain the vehicles because no one else in the area would work on them.

Most of the cars about town were Ford Model Ts, which customers could purchase for as little as $360. The Model T, known as the "flivver," "jalopy," and "Tin Lizzie," was changing American life. With the practice of the

The Dittlinger family's three Franklins were air-cooled vehicles, and therefore they had no radiators. Note that Franklin styling included a scoop-nosed front on these ca. 1919 models. The scoop nose was changed to a more conventional shape by 1922. (Photo by Joe Sanders.)

Joe Sanders displays the straight six-cylinder air-cooled engine of a Franklin. (Laurie P. Sanders Collection.)

World War era's "Motorless Sundays" over, people were once again driving for pleasure. In the early 1920s the touring car was the most popular. Many people preferred these more or less open cars, because drivers claimed the vehicles were less squeaky, less expensive, and not top-heavy. (Some closed cars could roll over and break apart.) Buyers thought open autos were safer, had more ventilation, and of course made "touring" easier. In fact, on March

14, 1921, the Comal County Commissioners Court ordered the purchase of a Ford touring car for county use.[16]

If "touring" was made easier, little else was. In an open car dust was a big problem. Drivers often wore long protective coats (dusters) and goggles. It was hot in summer and cold in winter. During a rainstorm occupants scrambled to put up the top—often a practice in futility, because by the time they completed the task the rain usually stopped. Then it got steamy. Cars did not have glass windows; they had snap-on curtains with small isinglass (transparent mica) windows sewn in. Car heaters were very rare. Some early heaters actually warmed the vehicle from the heat of the exhaust system, a potentially dangerous procedure that manufacturers soon abandoned.[17]

On Model Ts the gas tank was under the front seat, which had to be removed before a fill-up. Drivers crawled under the car to check the oil. They even tried quick fixes such as oatmeal or cornmeal to plug radiator leaks. Dealers sold automobiles as very stripped-down items. Modern essentials like bumpers, car horns, gas gauges, and speedometers were extras. And scads of catalogs advertised such accessory items for the discriminating driver. It was necessary to keep an emergency kit containing supplies like spark plugs, oil, light bulbs, fuses, and an assortment of tools for on-site repairs.[18]

One of the worst ordeals for motorists was getting the car started. Cars did not have self-starters. They had to be cranked, and a person could do a lot of cranking to get the motor running. If the crank slipped, the result could be dangerous. Many people suffered broken arms from a wild crank, which prompted companies to advertise all kinds of non-kick crank devices.[19]

Another big problem that affected all motorists was flat tires. Young Milton Kaderli, growing up on a ranch in the Comal County Hill Country, recalled, "Early on you always had to consider having about at least two flats coming into New Braunfels." San Antonio resident (who later moved to New Braunfels) Jarvis Hillje recalled his boyhood days of spending many Mondays asking a school friend about his family's weekend driving trip. The highlight of each story was always how many flats occurred and where.[20] Dorothy Wimberley Kerbow, whose father Calvin Hickman Wimberley was the water superintendent for New Braunfels, remembered their trips to Wimberley: "We had an average of three flat tires [every trip]. If we didn't have more than three flat tires my father thought that it was a real decent trip."

When a tire went flat on their Model T, her father gathered rocks to stabilize the wheels. He then jacked up the car, took the tire off, and pulled the tube out. The procedure for patching a tire, according to Dorothy Kerbow, went as follows:

> He had a little kit and it was called a "Cold Patch." He would take a little rough thing that looks something like you use to shred things with, a grater, and he would rub that spot where the hole was ... get it rough. Then he would take the paste. He'd tear it off, like opening a band-aid ... put this paste on it and put [a patch] on there and put a clamp on it, and hold it for a while. Then he'd put the tube back in the tire and pump a little bit of air in it with one of these hand pumps, and he would put it back on the wheel and pump it some more.[21]

Cars also overheated easily. As Kerbow commented, "Old Model T Fords were like old horses—they didn't like to pull a hill." By the time a vehicle reached the top, it inevitably boiled over. In fact, Model T owners often backed up hills because reverse gear provided more torque and more dependable power.[22]

As more auto dealers set up business, many developed creative sales gimmicks that would give them an advantage. In the early 1920s New Braunfels resident Egon Jarisch worked for Mauer Motor Sales selling the Star automobile, a competitor of the Model T. The town had more than just the original Ford dealership by this time.

> I had a peculiar way of selling that little Star automobile. I would take four good-sized men and myself and we drove ... the Klappenbach Hill

> [present Fredericksburg Road out of Landa Park].... I would stop right at the foot of the hill so the wheels, front wheels, would be just standing up on the grade. And I [would] start that automobile in high gear with those four men and myself in there.... I made that hill in high gear. Never changed gear, and when I got on top of the hill I was making 20 and 25 miles an hour.... That sold that automobile every time.[23]

The automobile business was picking up in Comal County. In the early 1920s Johnny Ruppel worked at nearby Gruene, Texas, for a Maxwell dealer, and at that time their sales were up to sixty-five or seventy cars a year.

Dittlinger, with increased confidence in Joe's abilities, depended more upon his driving. They became good companions and sometimes explored the roads together. He asked Joe to be chauffeur on an auto tour out West. Dittlinger, his wife Elise, and Joe set off on July 11, 1922, for Colorado, Utah, California, Arizona, and New Mexico, and returned in late October. In his travel diary Joe kept records of every place they saw including the conditions of roads along the way. Percentages of road inclines went as high as an eighteen percent upgrade for four miles followed by a similar steep

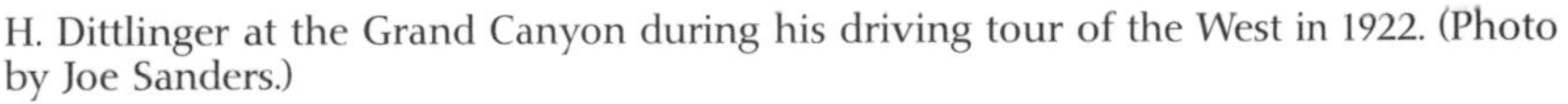
H. Dittlinger at the Grand Canyon during his driving tour of the West in 1922. (Photo by Joe Sanders.)

Joe proudly poses in front of his homemade driving machine–"Skeetzer." Built from spare metal and parts, the roadster sported wooden-spoke wheels, a cylindrical fuel tank behind the seat, and a large trunk attached on the back. (Photo by Laurie P. Sanders.)

downgrade for four miles on the way to Pinnacles National Monument in California. On September 9, 1922, they had a blowout while driving from Bakersfield to Los Angeles and purchased a new Goodyear cord tire. His entry on October 16, when they were en route to Winslow, Arizona, read: "Roads out of main good for a short way very bad from [Kingman] to Winslow projecting stones and chuck holes. Slow driving was required."[24]

After seeing the West and getting some of the wanderlust out of their systems, they returned to New Braunfels on October 24, 1922. It was back to his mechanical work at the mill and part-time chauffeuring.

Up until this time Joe did not have his very own car, so he decided to build one. From various scraps and parts he accumulated, he built a little two-seater that he named "Skeetzer." The roadster, basically a stripped down body, almost looked—if one stretched his imagination—like a snazzy Stutz Bearcat, the sports car of the day. Nevertheless, he tooled about town and onto the county backroads.[25]

He marveled at the scenic beauty of Comal County's green river valleys and rugged hills—a sharp contrast to Ottoville's flat farm country. Seeing the Guadalupe and Comal rivers, venturing to romantic-sounding places like Bear Creek and Hidden Valley, Joe developed a fascination for exploring the country byways and discovering what lay beyond the next turn in the road.

Heading for the hills! (Photo by Laurie P. Sanders.)

• • • • • Chapter Three • • • • •

Sweet Sunday Drives

The spring of 1923 was a time to work and explore and simply enjoy the pleasures the Texas Hill Country had to offer. Joe spent his bachelor days as a mechanic in the mill, and he swam in the nearby Comal River every morning before work. He also chauffeured Dittlinger about town. In his spare time he took interest in civic affairs and was active in the newly formed local chapter of the American Legion.[1] He explored the backroads in his Skeetzer and drove one of the Franklins on picnics with the Dittlinger family. As he became better acquainted with the beauties of the Guadalupe River, he tried to share his scenic appreciation with everyone he met by taking them on country drives. On Sundays he went to Landa Park (like everyone else) to watch the New Braunfels Tigers play baseball.[2] He didn't know how a chance meeting there would change his life.

Laurie Pfau was born in Clarksville, Ohio, but in 1905 when she was three years old the family moved back to her mother's native Texas, settling in the newly platted South Texas town of Sarita on the Kenedy Ranch. There with her three older brothers she spent her early childhood. While their father, Oscar Pfau, tried to make a go of it with his own general store, the children attended a one-room school, explored the scrub brush prairie, and met the train at the depot each day for that was the most exciting social activity. Travel of any distance was done in jolt wagons and proved to be a bumpy, cumbersome affair. Ranchers were very protective of their property and sometimes erected fences across the road. Consequently on one such occasion Oscar Pfau, armed with wire cutters, decided to make an impromptu trip down the road to the nearby town of Riviera. A trip to the bay, twelve miles, was an all-day event in which the whole town rode in wagons to go on a picnic by the seashore.

One man in town owned a car, and the contraption caused a lot of excitement when he drove around. The kids called the man "Uncle Billy," and they could both hear and see him coming from quite a distance—the motorcar made so much noise and smoke. It looked like the skeleton of a buggy minus the horses, and with all the trouble he had starting it and with the backfiring once the auto moved, a person could practically reach his destination faster on foot. Needless to say everyone thought "Uncle Billy" was a peculiar experimenter.[3]

The Pfau family left Sarita in 1910 and went to San Antonio in search of a better livelihood. Though they stayed only a few months, during that time they experienced firsthand the wonders of technology and transportation. The lighted streets and shop windows were an amazing contrast to the quiet, simple lifestyle in Sarita, and it took some time to get used to all the electric streetcars of the big city.

By June 1911 they moved to Victoria where Oscar went to work in another mercantile. Even in the larger towns the automobile was a new and unusual sight. The few cars that graced the streets of Victoria belonged to wealthy families. Some of those vehicles included electric cars and most of the motorists of electrics were women. Two female drivers commonly seen riding downtown were the Power sisters, neighbors of the Pfau family. The electric autos were quieter, cleaner, and easier to drive because they did not have to be cranked and did not produce billowing clouds of smoke.[4] One disadvantage, however, was the electric's limited traveling distance between rechargings.

Many of the roads around Victoria were muddy and deeply rutted. The road to Cuero was nothing more than a cattle trail, and drivers often needed the help of a farmer with a team of mules to pull them out of the mud.

About 1917 Lauric's oldest brother, Claude, became the first member of the family to buy a car. It was a Stoddard-Dayton, and changing a tire took forever because of the large size and weight of the auto. He sold it when he was drafted into the army during World War I. His father, Oscar, scoffed at the troublesome vehicle and held out for as long as he could before breaking down and getting an automobile. It would be 1926 before he would buy his first car—a Model T Ford touring car for $400. By then Ford would offer colors like Windsor Maroon and Channel Green, but Oscar's bare-bones vehicle was black and had a crank.[5]

After Laurie graduated from high school in 1919 she stayed in Victoria and helped her mother around the house. In the early 1920s she took occasional trips here and there—down to Houston and Galveston to visit friends. Travel near the vicinity of Houston included crossing the Brazos River by ferry, where Laurie suffered some anxious moments. The road was not topped, and the driver, a school friend and something of a show-off, sped to the muddy Brazos banks to meet the ferry. The river was in a flood stage, and Laurie held her breath when the car skidded in the red mud to a stop. The swift water lapped at the front tires, and the ferryman looked on with disapproval.

Then there was the excitement caused by Harvey Fite and Gene Cullum, friends of Laurie's, who hitchhiked from Houston to Victoria and on to Medina Lake west of San Antonio in the summer of 1922. Hitchhiking was so unheard of in those days that their adventure made the headlines.[6] But all in all, it was a peaceful, happy time when the latest fad of the day was to go "kodaking" on an outing. Every girl was equipped with a camera and a "uke" (ukulele).

In July 1923 Laurie Pfau planned a trip to New Braunfels. She'd heard about the area's wonderful scenery from her brother Karl who, while temporarily stationed at Fort Sam Houston in San Antonio, had visited Landa Park in 1918. She went to visit her old school friend Norma Braune (whose family had left Sarita and settled in New Braunfels), and Norma's new husband Philip Karrer.

Laurie traveled by train from Victoria to the station in San Antonio, where she met Norma and Philip. On arrival in New Braunfels she saw a

Bird's-eye view of downtown New Braunfels (ca. early 1920s) taken from the grain elevator at the Dittlinger flour mill. Market Street is clearly visible at the bottom of the picture, and East San Antonio Street crosses it. The large building on the lower left at one time housed the Comal Hotel Apartments and at another time operated as the Comal Sanitarium (and was Prince Solms Inn in the 1990s). Going up San Antonio Street, the Pfeuffer Home appears to the left, while beyond it on the corner of the town Plaza is the Eiband and Fischer store. Across the Plaza is the Comal County Courthouse (upper right) and to the lower left of it are seen the spires of the Landa Home. (Photo by Joe Sanders.)

clean little German town of about 3,500 people.[7] "When I came here visiting in 1923," Laurie recalled, "there were three things that everybody called your attention to that were new." Residents proudly pointed out the new Comal River bridge on San Antonio Street, the post office on Castell Street, and the Dittlinger office building on San Antonio Street.

She stayed with the Karrers in their apartment on San Antonio Street a few houses down from the Dittlinger mill and enjoyed strolling through downtown New Braunfels with her friend. Norma Braune Karrer had lived there for some time and described small-town life, especially the eligible bachelors. She talked about Joe Sanders and had even gone on a date with him the year before. He was the outsider, the northerner who possessed a mysterious quality that many local girls found attractive. One Saturday evening the two girls were walking past the mill by the Comal bridge. "Up comes a couple of fellows throwing a ball to each other," Laurie remembered. Norma nudged her and said, "That's Joe Sanders I was telling you about." But Laurie thought "So what," and they went on their way.

That Sunday they went to a baseball game in Landa Park. Laurie recalled:

> It was very boring since I don't care for baseball.... We didn't want to stay for the whole thing. When we were leaving the ball game, crossing a little bridge that goes over the Comal there outside of the ball park section [at the Landa Park entrance, Joe starts coming from the other direction, and Norma called him and said, "Joe come here. I want you to meet someone." So we introduced each other and ... he asked if we were walking, and I said, "Yes." He said, "Well, I've got a car.... I'll take you home."

Laurie Pfau stands on the new Comal bridge during her visit to New Braunfels, while the passing driver yells, "That'll be some picture!" The Comal River bridge was completed in 1923 by Miller-Fifield Construction Company of Waterloo, Iowa. (Photo by Norma Braune Karrer.)

Joe was driving one of the Franklins, a rust-colored sedan. He offered them a ride home, but first asked Laurie if she'd seen any surrounding countryside. She replied "No," and with that she, Norma, and Philip piled in. "We all got into the car, and he took us out driving into the hills, and I thought 'Oh how beautiful,'" Laurie commented. When they finally got home Norma invited Joe to stay for supper. After dinner they all went for another drive to the edge of San Antonio to see the skyline lights.

The time signified the start of a new life. Of the day, Joe simply wrote this in his diary sometime later:

> Sunday, July 22, 1923: Met Babe [her nickname] for the first time on the Bridge in Landa Park. At the time I had no idea she was to be my sweet little wife. Tho I loved her from the start, because she is the sweetest.

Joe extended an invitation to Laurie and the Karrers for a drive and a picnic the following Sunday. For Laurie it proved to be a memorable day as they motored through the Hill Country. "I'd never seen anything like that in all my life growing up in Victoria or Sarita," Laurie recounted. She marvelled at the beautiful scenery along the Guadalupe River and all the springs and ferns on the cliff sides.

They drove two cars. Joe and Laurie were in the Skeetzer, and Philip and Norma drove a Ford coupé. Everyone switched places periodically so that no one would have to be in the hot sun too long in the open Skeetzer.

"We went to a place he [Joe] had heard of called the Narrows," Laurie recalled. "I think he had seen it once before.... He wanted to see it again."

The Narrows was a beautiful spot on the Blanco River just across the county line in Hays County. In this scenic place, the river plunged down into a gorge and carved out a series of unusual terraces and potholes.[8]

"Well to get there we had to go through a farmer's barnyard," Laurie remembered. They stopped and asked the farmer's permission to visit the private park, and, with no road, drove cross-country to their destination. The cars bounced over rocks and across a meadow. They parked and the

Two chums, Laurie and Norma, stroll arm-in-arm down a Hill Country byway to a Guadalupe River crossing—probably Specht's Crossing near Spring Branch in west Comal County. (Photo by Joe Sanders.)

Joe peers down the chiseled cliffs along the Blanco River. Ferns are growing out of the side of the spring-fed rocks. (Photo by Laurie P. Sanders.)

foursome walked the rest of the way to the edge of a narrow ravine. "It was so narrow that it looked like you could almost jump across it," Laurie described, "but of course you couldn't." They looked down upon a crystal-clear river that trickled over rocks upstream and suddenly cascaded into a deep limestone corridor. The cliff sides were lush with maidenhair fern, and the constant dripping of the springs in the cool and shadowy place possessed a pleasant, almost musical quality. Downstream, the picnickers negotiated their way over the cliffs and down into the heart of the stony banks and the river's edge. Laurie slid on the slick rocks and soaked her shoe, but Joe caught and steadied her. They stayed at the Narrows until mid-afternoon, then drove through the Guadalupe River Valley. By sundown they were back in New Braunfels.

A few days later Laurie had a date with Joe, and they went for another ride. She recognized the unique beauty of the Hill Country. The clear green rivers and rugged cliffs marked a sharp difference from the flat land and

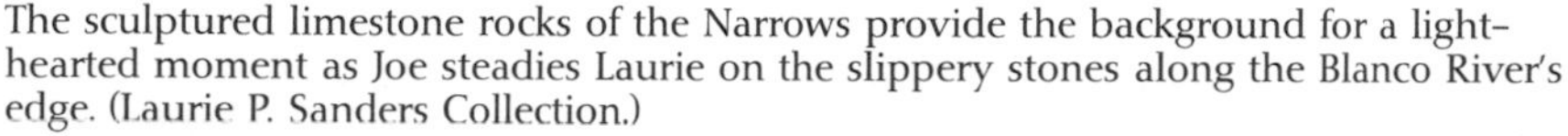

The sculptured limestone rocks of the Narrows provide the background for a light-hearted moment as Joe steadies Laurie on the slippery stones along the Blanco River's edge. (Laurie P. Sanders Collection.)

July 29, 1923—Joe and Laurie at the Narrows. (Photo by Norma Braune Karrer.)

muddy Guadalupe down in Victoria. She really loved the area and felt a sadness at the thought of leaving, but soon her three-week trip came to an end, and early on Sunday morning Joe and the Karrers took her to the train station in San Antonio.

"He got on the train with me," she remembered, "kissed me goodbye, and said that he was going to come down to Victoria to see me." With time and distance between them, she didn't really believe him, but he kept his promise and visited her every month. They were married the following spring, and Laurie went to live in New Braunfels.

Their first residence was an apartment on Jahn Street. After a short time they rented a house on Mill Street for $15 a month. Located several blocks west of the railroad tracks, their house was the last one before the open country. Beyond was a large cornfield and vast expanse to the nearby hills. Laurie said, "You could look all the way up to the edge of the hills, and at night you could hear the coyotes. If you've never heard a coyote you'd know one when you hear him. You could hear them [howling] at night."

Laurie tried to adjust to life in the little community. New Braunfels was very much a town of order and tradition. Most families were descendants of the original German settlers from more than seventy-five years earlier, and in many areas German, not English, was the preferred language. Outsiders had to adapt to the life and culture, and it was not always an easy adjustment. Joe could speak German fairly well and got along with many a merchant, laborer, and rancher. Laurie, not speaking German, had a tougher time adapting to the old world customs that pervaded almost every store and neighborhood.

The first real influx of *Ausländers* occurred about the time Laurie settled in New Braunfels, when the power plant was under construction. The U.G.I. Contracting Company of Philadelphia directed the work in 1925, and a group of engineers, carpenters, and other workers came from the North to build the plant on forty-five acres next to Landa Park. The towering power plant would become a familiar landmark in the years to come, but its construction also represented early signs of change. Workmen brought their entire families down with them, and some bachelors married local girls and stayed on afterward. The influx accounted for a temporary shortage of housing in New Braunfels, greatly affecting the Sanders' prospects of finding a place to rent.[9]

As a young wife setting up housekeeping, Laurie also had to familiarize herself with the daily practices and routines at home. The households about town stayed neat and clean, making the most of what land was available. Many families grew vegetable gardens in their front yards. It was apparent that Mondays were set aside for washdays as women scrubbed and hung clothes and aired mattresses outside. Because all the clothes wrinkled, Tuesdays, of course, were ironing days.

An iceman came around every day, and people displayed a placard in their front window with the numbers 10, 15, 25 and sometimes even 100 on it, representing the pounds of ice that the household wanted. Dorothy Wimberley Kerbow grew up in New Braunfels in the 1920s and remembered that her daily chores included emptying the drip pan full of water

A prominent landmark in New Braunfels—the power plant officially opened on November 6, 1926, and was called the Comal Power Company, an associate of the San Antonio Public Service Company. The postcard dates from the late 1930s. (Photo by Otto Seidel.)

that was kept under the icebox. She recalled that the iceman was always a welcome sight. "He was in a wagon with a mule pulling it." The mule walked slowly while the iceman looked ahead at the next placard hanging in a nearby window and cut a chunk of ice—on a bed of straw and covered by burlap—and was usually exactly right. He carried the ice with big tongs. Kids would follow behind and eat ice chips.[10]

If the iceman was a common sight, the trashman certainly was not. There was no garbage pickup in town. Households usually kept a few chickens to eat up leavings, and everyone made compost piles. Periodically, they burned anything that was left over.

On Saturdays farmers still came to town in their wagons and stayed the night. Laurie remembered seeing many wagons parked in the lot by Eiband and Fischer's store at the downtown Plaza. For many of them the journey from the hills was too long and cumbersome to attempt a round trip in one day. Besides that, the weekly trip was as much a social ritual where farmers and townspeople got together to catch up on the latest news and talk about the day's trip. The day ended in festivity with a dance somewhere on Saturday nights.[11]

The country roads were rough, but travel in general was slowly improving. At least on the highways most everyone had access to transportation through new bus services. Austin and San Antonio papers advertised the Red Ball bus lines from San Antonio to Austin.[12] For a couple of dollars a traveler could ride the bus to Austin or San Antonio through New Braunfels and San Marcos and back. The early buses, however, were nothing more than touring cars that could carry six passengers plus the driver. "That was the Greyhound bus of the 1920s," Dorothy Kerbow recalled. The vehicle flew two pennants that displayed a red ball. Along the Post Road, the vehicle could travel only twenty-five to thirty miles per hour on the good sections; slower on the rougher parts. There were no rest stops along the way. According to Kerbow, "When they'd stop for a rest stop … everybody had to go for the bushes." About the mid-1920s a few tourist courts operated along the main highways and served as rest stops. They were the

The country backroads beckoned to adventurers out for a quiet, scenic drive. (Photo by Joe Sanders.)

primitive motels of the day. The tourist court consisted of a few cabins, a store, and an outhouse and shower for the whole establishment. Travelers stayed for $1 a night. As bus services increased and passenger demand grew, companies exchanged some of the touring cars for twelve-passenger split vehicles that had their frames cut out to make larger cars—"stretch jobs" they were called.[13]

By the mid-1920s Highway No. 2 was under the control of the state highway department. The state imposed a one-cent-per-gallon gasoline tax in 1923. In 1924 the Texas Highway Department took responsibility for maintaining state roads and in 1925 also took control of constructing new highways. Counties, however, shared financial responsibilities and still did some maintenance with county-owned equipment for a few more years. That year 983,420 vehicles were registered in Texas resulting in the consumption of 440,618,929 gallons of gasoline.[14]

In Texas and Comal County the foothold of the motorcar was forever established, but many of the roads were still wild and woolly. Scenery lovers traveling the Hill Country, however, kept keen their sense of adventure.

• • • • Chapter Four • • • •

Cattle Guards and Bumper Gates

These are the days of the tour and to thousands more each year it means a trip by automobile to lake or mountain, or the more pretentious visit to the old home, or to some new, and to the tourist, undiscovered, country. The tourist who goes by auto fully prepared for the emergencies which may arise, unhampered by railway schedule, and who will take time to enjoy himself, will get untold pleasure out of the trip.

—*Putnam's Automobile Handbook*, 1918

Motorcar books were full of advice for touring drivers, who took all the comments to heart. Manuals recommended that motorists carry along an assortment of tools even for the shortest spin. Extra inner tubes, spark plugs, electric bulbs, nuts and screws, oil, and a cold-patch kit were absolute necessities, as well as an ax to chop away branches obstructing the roadway.[1] Warnings about driving courtesy were just as important and included earnest reminders to heed "that the plodding farmer's wagon has the same rights to the road as yourself, and that horses and carriages were using the highways long years before motor cars were dreamed of."[2]

It was the journey, with all the struggles involved, that determined the joys of the adventure:

The speedster knows nothing of the pleasure of touring, and the fellow who is always worrying lest he shall not make Squedunk Corners in time to

> eat and get to Possum Crossing for the night will be watching the road map and the clock on the dash so closely that the beauties nature has lavishly scattered about will be lost upon him, and his memory of the trip will be a procession of eating and sleeping houses, with trouble between them all.[3]

Joe Sanders was certainly one who appreciated the excitement of the drive itself. He was an expert mechanic who was confident in his abilities and had all the tools and knowledge needed to enjoy a tour of the countryside. He followed many lonely backroads, thereby discovering the beautiful views the Hill Country had to offer. On warm evenings and Sunday afternoons Joe and Laurie drove out in the Skeetzer. When going for a scenic drive, they started out from town past Landa Park and up Klappenbach Hill. Once up to the summit they turned onto the road to Blanco and headed for the hills.[4]

"We'd go to Startz Hill. That was one of the high spots in the county.... You could sit and look way down onto a valley," Laurie said.

The Skeetzer had been a fun roadster for its time. The car was not practical, however, for anything but good weather. By the later 1920s the Sanders moved from their humble home on Mill Street to South Street and finally settled on Washington Street on the other side of town across the Comal River in an area known as Comaltown. By now their household included a third member, their daughter Laurie Jo. It was awkward to drive about in the open Skeetzer with a baby in the front seat. At first Joe fashioned a top for the car, but eventually he realized it was time for another automobile. About 1927 (the last year of the Model T) he acquired a Ford Model T coupé that had been stored in the shop at Dittlinger's mill. The black enclosed vehicle had one long front seat, with gas tank underneath. Every time Joe filled the tank, Laurie had to get out. But the auto proved adequate for their needs, and, in addition, Joe usually had access to other mill cars. By the late 1920s closed-type vehicles were more popular than many of the open touring cars, and nine out of ten autos manufactured were enclosed.[5]

In the coupé the Sanders family resumed their outings. The wonders of mechanical contraptions fascinated little Laurie Jo at a very early age. As a toddler she played with toy cars as well as dolls, and, seeing her father's love for autos, even referred to herself as "The Car-Car." She had a little toy black coupé of her own and made "motor" sounds as she played. But as a tot, she was slow to appreciate the adventures of driving the backroads. Instead, she preferred the Saturday night window-shopping on San Antonio Street in downtown New Braunfels—a popular practice that many families enjoyed. When faced with the prospect of touring, she inevitably protested, "You don't want to go to the countryside!" Upon reaching the outskirts of town she asked, "Is THIS the countryside?"[6]

Their drives took them all over the county and beyond—out to the north near Sattler, atop the Devil's Backbone, or along the Guadalupe River. The hamlet of Gruene about three miles northeast of downtown New Braunfels was also a common destination. It was a quiet place of scattered homes, a cotton gin, and Gruene Hall, where they occasionally stopped for cold drinks and hamburgers. The Gruene low-water bridge, constructed over the Guadalupe in 1910, was covered with water some of the time.[7]

Sometimes the Sanders family took a spin on the Post Road toward Austin. About five miles up the road was a long incline known as Austin Hill. From that vantage point viewers could look back upon the town of New Braunfels and the whole valley. The section of the Post Road between Gruene and the village of Hunter was especially enjoyable to the little Laurie who squealed with delight while swooping down through all the sharp dips in the road.[8] By the late 1920s officials planned even more work on the Post Road in order to cut more distance off the route between Austin and San Antonio by straightening dangerous corners and curves.

A favorite picnicking spot for the family was Bear Creek, one of those pristine, undisturbed places with water, rocks, and greenery. Bear Creek Road branched off the New Braunfels-Blanco Road (Highway 46) and traversed what was regarded as some of the wildest, most rugged territory in the county. About eight miles down the road were the old signs of a

Bear Creek's honeycombed cliffs were a popular destination for scenery lovers. (Photo by Joe Sanders.)

permanent Indian camp where, in ages past, tribes kept their fires perpetually burning. A huge charcoal bank covered an area from fifty to a hundred feet in diameter and contained Indian relics of tools and arrowheads.[9]

Bear Creek was a spring-fed stream that carved out a picturesque, honeycombed valley. East and west forks came together to form one creek. It was this 4,800-acre wild area that the Dittlingers, along with several other prominent New Braunfels families, purchased in 1921. Originally used for hunting club purposes, by the late 1920s the club folded and the owners designated the land as a nature preserve.[10] As Dittlinger grandson Hip Mengden recalled, "It was natural in that it had never been developed, and you just leave it the way it is. You come in … and go out leaving only footprints and tire tracks, and so it was."

Because Joe was an employee and friend of the Dittlingers, he had permission to visit the preserve and had a key to the gate. Laurie would pack a lunch, and they picnicked there often, especially in the summer, for the scenic spot was always shady and cool. Bear Creek, with its ever-flowing shallow and deep rivulets over the textured limestone, was a special place.

Laurie remembered the sheer fun of just looking around. They found all sorts of arrowheads on the cliffs above the creek, and the side walls of the cliffs were covered completely with ferns. In one small grotto, they could cup their hands under a seep and fill them with pure spring water. "We would drink that water," Laurie said. "It was good, clean pure water.... It was shady ... and you could just sit there in the coolness.... The beauty of it was untouched."

Daughter Laurie Jo even as a very little girl was impressed with the natural beauty of Bear Creek. "We used to wade in the water there," she reminisced. "It was just a nice depth where you could wade, and I guess there were deeper regions where you could swim if you wanted to.... It was a nice place to picnic."[11]

These family picnics set the stage for many outings to come. And likewise many people at the time were beginning to enjoy a scenic day-trip. In April 1928 the *San Antonio Express* published its own road log for a country

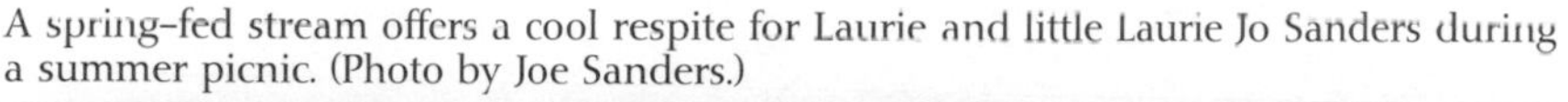
A spring-fed stream offers a cool respite for Laurie and little Laurie Jo Sanders during a summer picnic. (Photo by Joe Sanders.)

tour. That excursion took the driver on a "12-hour loop trip from San Antonio through Castroville, Leakey and Kerrville." The log started with advice that the journey "should only be attempted after dry weather and on a clear day with no prospect of rain." From mile point to mile point it contained many interesting and important warnings and comments typical of the day.[12]

On any trip it was important to get an early start and have plenty of gasoline. At that time most general stores and automobile garages sold gasoline. (Joe usually secured fuel at the mill.) The large "visible supply" glass gasoline pumps that held five to ten gallons in their upper glass reservoir had come into use. The fuel ran from the upper tank through the pump by gravity. The customer could watch the gasoline drain from the large glass cylinder, and by visible measurement markers, the attendant could determine the number of gallons sold. "And it took about ten strokes to get about a half a gallon," Hip Mengden recalled. The price ran from about 15 to 18 cents per gallon.[13]

Once on the tour travelers had to be prepared for numerous inconveniences. The road log continued with a barrage of descriptions like:

> … no road signs…. Road rougher, badly worn and frequent holes…. Ford river…. Fords, all on concrete floor. Remember brakes do not take hold so well when wet…. Just a rocky trail through the woods with occasional open pastures. Have to stop many times to open and close gates…. Enter ranch paddock, look out for the bull. Better blow horn and get someone from ranch to open gate…. Bump gates….

Of course not all of the roads were rough, but on a country byway and depending on the weather one had to be cautious. Sometimes getting an eyeful of reward took work like traversing many farms and ranches and encountering cattle guards and gates along the way. It was common courtesy to close the gate behind you, but that meant many a stop and start. Bumper gates helped alleviate that problem. They were, as the name implies, large swinging gates that a car bumped with its front bumper.[14]

William Kolodzie, a Comal County resident since the 1940s and later a county surveyor, explained. "You just hit them with the bumper of the car and they'd have a pendulum and they'd swing...." The motorist always drove on the right-hand side of the gate going in and the right-hand side going back. The gate always swung counterclockwise.

The gate, controlled by a large pole in the center with a heavy cable running through it, would swing back from its own weight. But a driver had to use skill in bumping the opening and time his entry lest the gate swing back and hit the car.[15]

Cattle guards and bumper gates, river fords, and mudholes were a part of the driving life. They underscored the adventure and made the prize of beautiful vistas all the sweeter. For the Sanders family a good drive was a breath of fresh air, a cool drink, and a time to enjoy, but one outing in particular brought to light the fact that they did not rule the road, but the road ruled them.

One Sunday in the summer of 1928 Joe decided to take a short joyride up in the hills. It was between ten and eleven in the morning. Laurie had put a roast in the oven at low heat. Then they called three-year-old Laurie Jo and Claudine, her seven-year-old visiting cousin, from their play to come along on the jaunt. The family ventured out in the black coupé down toward the Guadalupe River Drive. Joe motored through, crossing the fords of the river, while Laurie held her breath hoping that he was sure of the precise spot to cross safely. As they meandered along, Joe in search of this and that road, everyone realized that they were lost. They rambled out toward Sattler with its wild valleys. It seemed to Laurie that just when she thought all the crossings were behind them, they'd come upon another river ford. It was lonely, rough country, and there were no houses in sight where they could stop to inquire about directions. Joe felt that he'd gone too far to turn around and backtrack. He was convinced that if he searched long enough, he'd happen upon a familiar route that looped back to home. But Laurie and the children were getting scared. The cousin Claudine, accustomed to

Victoria's flat land, was frightened by the continuous steep hills. Little Laurie, scared about the water crossings, just wanted to go home. Her mother worried about the Sunday dinner.

They drove on endlessly—Joe determined to find a way and the rest objecting and wanting to quit the exploration and find the quickest path home. At last he came upon a road that looked familiar and figured out the way back to town. It was about three o'clock in the afternoon when they finally made it—their five-hour adventure was a far cry from a quick half-hour spin.

The backroads were more intricate and isolated than Joe had bargained for that day, and the experience taught him a valuable lesson about being prepared and respecting the road. Laurie, on the other hand, worn out by their adventure, was relieved to discover that Sunday dinner was intact. It was the best roast they ever had.[16]

• • • • Chapter Five • • • •

Signs of the Times: Travelers, Bums, and Entrepreneurs

As the 1920s drew to a close, public support for better roads in Texas grew ever stronger. In 1929 Texas boasted 18,728 miles of main highways, of which 9,271 were hard-surfaced. However, a staggering 169,836 miles of rural roads existed statewide, and only 13,846 of those miles had any kind of hard surface. Surprisingly, of the total 188,564 miles, there was not one continuously paved highway in the state.[1]

With the increased traffic, the Texas State Highway Department focused on improvements to major thoroughfares, including plans for a new highway for the San Antonio-Austin Post Road. The new course would straighten the route between the cities, cutting off eight miles. The goal to eliminate dangerous curves and railway crossings, though desirable, presented a dilemma for the towns of New Braunfels and San Marcos, which feared the new way might bypass them. The department attempted to provide a solution by planning a new straighter course to go through the two towns. The proposed new highway was to have pavement forty feet wide and a total width of a hundred feet.

San Antonio and South Texas claimed the best roads in the state. Travelers espoused an awareness and enthusiasm for the need for good byways, and the supply of limestone in the area provided abundant material for more road building. In 1929 the *San Antonio Express* boasted that

Bexar County had the "largest mileage of improved roads of any County in the United States."[2] It also stated that because of the "splendid highways over South Texas more than 150 motor busses operate from San Antonio daily to all parts of this section, as well as into other States...."

The increased mobility enabled people to see and experience new and different places. Bus lines took some of the credit—a message made clear by a large feature on motorcoach transportation in the October 29, 1929, issue of the *San Antonio Express*:

> The motorbus has brought about a social revolution in all civilized countries. It has removed the isolation of even the most out-of-the-way places. Thousands of years ago the Greek poet Homer ascribed the greatness of Ulysses to the fact that he knew "the ways of many men and many cities." It is exactly in this respect that the motorbus is a great liberalizing influence. Before its advent, numbers of men and women were born and died without journeying 40 miles from their birth places. The narrowness of their physical horizon produced a narrowness of mental outlook resulting in prejudice, misunderstanding, distrust. The development of the motorbus helped to change all this; it has made the distant farm or village the suburb of the city. Travel is a great educator and a great civilizer, and the motorbus has made travel easy.

The same *Express* issue remarked on the convenience of the San Antonio area for exploring motorists. Many visitors on their annual trips to South Texas made San Antonio their headquarters from which to embark on tours through the surrounding territory.

Many drivers headed for the backroads of Comal County to admire the rugged Hill Country scenery. The Sanders family traveled every Sunday and on some free weekday afternoons and evenings. They often met out-of-town tourists, especially on Sundays. Roads along the Guadalupe River or Startz Hill that had previously known only the local traffic of farmers and ranchers now saw an increasing number of motorists out for a spin and a picnic. The maze of the unmarked country backroads presented a confusing dilemma for lost drivers. Almost every week some tourist along the way stopped Joe to ask for directions. He was happy to oblige. Joe had become more experienced at driving a strange byway until familiar territory appeared again, because he had spent the last couple of years discovering the intricate network of country roads. His knowledge of the various turns, hills, and river fords gave him more confidence to explore.

In addition to touring the Texas Hill Country he journeyed to other places, taking an occasional jaunt across the state. Joe and Laurie had traveled to Glen Rose southwest of Dallas in August 1929 and had stopped to see the dinosaur tracks in the Paluxy River. The trip consisted of driving a series of major and minor highways, but true to the current highway statistics, there was not one continuously paved route. Instead, they traversed a patchwork of dirt and gravel roads.[3]

In the fall of 1929, as well as his mechanical work at the mill and chauffeuring duties, Joe focused on civic activities. In his spare time he challenged his skills by building an "airplane"—at least the contraption looked like an

Joe Sanders leads the Armistice Day parade through downtown New Braunfels in his "Aeroplane" in 1929. The plane was covered with wire mesh and paper poppies. For the next few years, Joe drove the plane in other parades in the area and across the state as a unique float advertising Comal Post 179 of the American Legion. (Laurie P. Sanders Collection.)

airplane. But it was not intended to fly. Joe could only drive it, which he did as the American Legion float in the Armistice Day parade. The plane, covered in red poppies, was quite a spectacle leading the procession with Joe at the controls. After the celebration, the plane was on display at the Comal County fairgrounds, where poppies were sold for about a quarter each. Money benefited various levels of the American Legion, and, occasionally, hidden among the poppies, were $1, $5, and even a few $10 bills—added incentive to purchase a lot of flowers in the hope of getting a cash bonus.[4]

The following spring, with the warmer weather, the family took to picnicking and scenic driving again on a regular basis. One warm Sunday they met up with a lost motorist. He was alone and stopped to ask for directions. He introduced himself as A. W. Grant, the managing editor of the *San Antonio Express*. Grant was interested in the beautiful scenery but found the byways hard to navigate. He asked why there were no road signs to go by. Joe introduced himself and his family, chatted a bit, and gave him directions to wend his way back to civilization. But as Grant drove on his way, he had already planted the seed in Joe's mind. Joe acknowledged that there were no road markers and that the family routinely stopped to help a number of lost tourists. It would indeed be a great help to have signs to follow and would be a very worthy undertaking for somebody.[5]

As time passed, the county roads saw more than just Sunday drivers and scenery lovers; the region began to notice the effects of the Depression. People, down and out, traveled across the country in search of work. Laurie Sanders recalled "whole families packing up everything in their tumble-down cars and going from one part of the country to another." The desperate times sometimes led to criminal acts. On one backroad while returning from San Antonio, the family happened upon one unfortunate victim. The encounter had a vivid impact on five-year-old Laurie.

"We saw this man sitting on a suitcase beside the road," she later remembered. "He had been held up, and the people had taken his car, taken his money, and left him there with a suitcase full of stuff."

The man was a traveling salesman, with nothing left but his wares to display. He burst into tears while telling his story and asked for a ride into town. Joe drove him on to the nearest hamlet.[6]

More and more people were out of work. "People were desperate to get along," Laurie recalled. "Most people, if they had a job they lost it, or their pay was reduced ... so you just had to make do with what you had. He [Joe] was lucky in that his pay was not reduced." Joe's modest wages of $90 a month may have seemed like a princely sum to the many jobless. Newspaper ads were filled with workers looking for any employment. Families even sent their children peddling goods door to door. The Sanders household regularly had callers such as young girls selling packages of needles, pencils, cards—anything to make a few extra pennies. Sometimes men stopped by to offer to perform chores in exchange for sandwiches.[7]

Most hard-luck cases were genuine, but people sometimes took advantage of the hospitality. Joe helped out one elderly man who had been a gardener in California. He took the down-and-out fellow to a farmer friend who gave the man a job, some chickens to tend, and lodging in a small shack on the edge of the farm. The Sanders family visited the man occasionally, taking food. The farmer discovered, however, that one day the man literally flew the coop, stole the chickens, and sold them at Eiband and Fischer's store on his way out of town.

For the most part, though, people cooperated with each other. "You didn't have much yourself," Laurie commented, "but you were willing to help the other fellow who had less.... While it's a tragedy in one way, it brings out the best in other ways. You're willing to help each other when you're all in distress...."

Civic groups like the Lions Club gave food to the needy. Transients received a meal and a place to sleep in the county jail. The Dittlinger mill put up flour in sacks that customers could sew into little dresses, and Mrs. Dittlinger became renowned for giving away free pairs of shoes. It seemed that people from all over the country had heard of the lady in New Braunfels who gave away shoes.[8]

Besides riding in rundown motorcars, many transients traveled by freight train. When the trains rolled past the depot, it was a common sight to see several men (and women) riding the rails. Once after having gone to a movie, the Sanders family returned home to find three "bums" camped out on their front lawn. They had traveled in boxcars from Joe's hometown of Ottoville, Ohio, and hoped that their luck might improve in Texas. The

Dittlinger mill truck in the Comal County Fair parade (ca. early 1930s). Little girls sport their pastel gingham dresses and matching sunbonnets sewn from Dittlinger flour sacks in these scrapbook pictures. Laurie Jo is third from the left, bottom photo. (Laurie P. Sanders Collection.)

situation was awkward. Joe wanted to help out fellow Ohioans, but Laurie was concerned about taking in three strangers. Finally, Joe rented a room for them at the Comal Hotel in town.[9]

Wishing to make the travelers feel more at home, the next day Joe offered to take them for a scenic drive up in the hills. Laurie packed a large picnic basket full of sandwiches and fruit, and in one of the Dittlinger mill cars, the group set out with the family in front and the men in the back seat with the picnic basket. As Joe drove through picturesque valleys down the winding roads, Laurie noticed that the fellows seemed more interested in the sandwiches. Apparently, even the scenery was not enough to encourage the three to stay in the area, and with little prospect of work, they accepted some traveling money from Joe, hopped a freight train the next morning and headed west.

In spite of the large number of indigent people passing through, the citizens of Comal County did their best to help others and provide for themselves. Milton Kaderli recalled his boyhood observations of the hard work and self-sufficiency that characterized many families at the time. "There were lots of hardships, but the people could make the best of them," he said. "Everybody more or less didn't have any money, but they had enough somehow or other that they were able to get by." Families that owned land supplemented their needs with crops such as oats, wheat, and cotton, and they raised cattle and chickens. In addition to livestock, venison was an important food source. The Kaderlis grew much of their food at their ranch. "Mother had a beautiful garden, and she would just plant everything out there that you could imagine. We would can stuff, and we had a cellar [that] was always chock-full," Milton recalled. "Mother had chickens. We would have eggs to sell. She had cows to milk, and made butter," he added. "In times of the early '30s, we'd come to town and bring these cases of eggs and butter that we would sell to stores...."[10]

Others around the state may have practiced more cunning methods to make some money. By this time travel by auto had become a necessity for many people, and some exploited that need by trying to capitalize on road troubles. The journey from Houston exhibited distinct examples of underhanded savvy, as experienced by Hip Mengden, when he accompanied his father on drives to visit his grandfather Dittlinger.

> I can remember coming from Houston to New Braunfels in the summer.... We'd leave Houston five o'clock in the morning, and my father always wanted to make [it] to Gonzales for supper. There's a hotel there where we could stop.... And we wouldn't get to New Braunfels until 10 or 11 [P.M.]. Now that includes two tows, 'cause there were two places you could not get through. One was at Plum Creek out there between Seguin and Gonzales, and the other one was between Weimar and Schulenburg. Well the road was so wet and muddy in the middle of summer, even if you hit it at fifty miles an hour, you'd bog down, and then came the farmer with his two mules and for five dollars you'd get pulled out.... It cost you five dollars a time, ten dollars every trip. Oh, he watered it down every morning! We knew so! He spent all night hauling the water from the creek up on the road. My Dad would say "I'm gonna make it this time!" And boy, he'd just hit it, and we'd get almost across, and that was it!

Most stretches between New Braunfels and Houston were not paved, and if getting stuck in the mud wasn't enough, flat tires were also a problem. Normally Hip and his father had two flats on the trip. After Mr. Mengden put a spare on, he would buy another one in the next little town. The tire business could be profitable, especially if someone put nails on the road.

Hip explained:

> Well it was during the Depression and anyone who could make a buck was doing it.... These were "entrepreneurs...." I was always asking, "What's the name of this town?" And one of the towns between Houston and San Antonio is Flatonia, and so I said, "What's the name of this town?" "Flatonia." And I looked out there, and there's two tires with two flats. I says, "I know why they named it!"[11]

The state highway department did what it could to aid motorists in their efforts to prevent flat tires. About 1929 the department had put into service a magnetic nail picker and was the first in the nation to do so. The contraption consisted of a truck carrying a generator that powered a large electro-magnetic bar that ran underneath. The magnet was approximately eight feet long and eighteen inches wide and could be lowered to less than four inches above the ground. The homemade rig putted along at about eight miles per hour and could be quite effective. The highway department operated three nail pickers in 1930, and one report listed 6,062 pounds of nails and metal gathered in twelve counties over two years.[12]

Many troubles of the time were characterized in motoring terms. One item in the *San Antonio Express* summed it up in its "Headlights For Lightheads" column:

> Happiness must be cranked up, but troubles all have self starters.[13]

In attempts to "crank up" better times, New Braunfels citizens devised ways to improve public morale and the economy. Joe was still perfecting his road-marking idea and thought it would be an asset to the area. The minutes from the American Legion meeting on January 14, 1931, mentioned the idea:

> The first subject to come up for discussion was one mentioned by J J Sanders and was that of marking our leading highways and scenic drives. After some discussion, Comrade [Tom] Hughes [m]oved that the Commander appoint a committee to work in connection with the County and the City Board of Development with regards to securing the necessary markers. This motion was seconded by Comrade [J. M.] Francis and carried. On this committee, the Commander appointed Comrades, Sanders, Francis, Arthur Schumann and [Bob] Tays.[14]

At the next meeting on February 5, 1931, the committee on road markers reported that it had met with the county commissioners but that nothing definite had been decided. Evidently, no one was quite ready for such an endeavor, for there was no further mention of the task. Still, Joe didn't feel thwarted in his efforts to promote the Hill Country. On his own he planned and conducted scenic country drives for various civic groups.

Other organizations tried their hand at community encouragement. In the spring of 1931 the Lions Club sponsored a contest for a city slogan, calling for a "snappy and peppy" phrase that would stand for "what New Braunfels represents." They advised, "Everybody is invited to participate in this contest. So put on your thinking cap."[15] Lions Club members suggested some slogans to serve as helpful samples to contestants. Some phrases included "Where the Moonshine Spends the Summer," "Where Mother Nature Finished Her Masterpiece," "New Braunfels, the City that Attracts Visitors," and "The Oasis of Texas." In the following weeks the local newspaper printed some of the more colorful entries: "The Progressive City," "The Home of Good Character," "The Purest of the State," and "Beauty Queen of the Southwest."

Laurie remembered the contest:

> Joe and I were talking about it, and I said, "Well, why don't you say something like 'The City You Can't Forget?'" I just thought that that was a good slogan. You come to New Braunfels, you'd be enchanted by it as I was and he was. You just couldn't forget it. You'd always think about going back to that city you know. I thought that would be nice, and he said … he'd turn it in.

The next thing they knew was that the Lions Club chose the Sanders' slogan as the winner. Laurie, not wanting to make a public appearance, passed off the responsibility and gave credit to her husband.

> Joe Sanders Wins Lions Slogan Offer
>
> Judges of the slogan contest sponsored by the Lions Club have come to an agreement, and announce that Joe J. Sanders is the winner of the

> $10 offered for the most appropriate slogan for New Braunfels. Mr. Sanders' slogan is "The City You Can't Forget."
>
> According to the expressed wish of Mr. Sanders, the prize was presented to the American Legion Drum and Bugle Corps at the Lions luncheon Thursday.
>
> Judges in this contest, which was pronounced successful from every standpoint, were R. H. Wagenfuehr, Harv[e]y Richards, and Dr. R. C. Reynolds.
>
> The winning slogan will be passed to the city commission and City Board of Development, who will discuss it and decide whether or not it will be officially adopted.[16]

Apparently, the City Board of Development never adopted the slogan. In the interest of advertising they decided to adopt a phrase that had been informally used in the past—"The Beauty Spot of Texas," which had already been printed on county license plates.[17] Along with these moves to encourage tourism and give the town some notability, Joe finally decided to act upon his idea to mark the county roads.

"He got the idea that he would log the roads and get the different distances from one point to another and find out where they went and take all that information down and see about making some signs," Laurie said.

The undertaking took time and patience as Joe and Laurie set out week after week, Sundays and evenings, during the summer of 1931.[18] "While he would drive and keep note on the odometer how many miles from this point to that crossing, I would write it up in the notebook and keep track of it. We got to learn the whole territory up there," Laurie recalled. They spent the first months entirely on logging all the roads. It soon became obvious that, in addition to road signs, it would be beneficial to make a county map showing the network of winding byways. At that time there were no maps of all the county backroads, much less ones that marked scenic attractions. Many maps were generally sketchy.[19] The state highway department did print a map of the road system in 1927, though it was "made up of a series of straight lines between cities and did not show actual highway location or distances."[20] There were only geological survey maps of Comal County from the early 1920s. These contained accurate elevations and terrain features, but large areas were barren of road markings. One map of a 1927 section near Smithson's Valley listed only "many private roads through this section which is covered with cedar."[21]

In order to get as accurate a distance reading as possible, Joe followed a road and traveled it three times. He then compared the area to geological survey maps. Roads included county routes on ranches, across streams, and down private drives. Many private roads and parks required asking an owner's permission to stop or pass through. As a friendly talkative person, he easily communicated with many of the country people and was usually welcome on their land. He became acquainted with many of the ranchers in the hills. It could be a tricky situation while Prohibition was still on. Some farmers had stashes of homemade beer and wine and didn't like strangers poking around. Even so, Joe got along with most everyone. People were generally more trusting then and took a man at his word.[22]

Joe and the family did most of the road logging in a newer car—a 1931 Ford Model A that Joe "inherited" from the mill. It was black with balloon tires, large wide tires that were designed to provide a smoother, more comfortable ride. The Model A at that time sold for about $435, and of the total of fifty million autos made in the United States by 1931, Ford had produced twenty million.[23] Joe had gotten rid of the coupé, and Laurie was learning to drive the Model A in order to take their daughter to school the following fall. The Model T, though supposedly outdated by this time, was still heavily used by citizens saving their dollars. An old Model T sold for as cheap as $15, and Laurie's mother, on various visits to Comal County, often commented on the large number of "old cars" still driving on the streets of New Braunfels.[24]

In addition to driving with the family, Joe also logged the roads with Mr. Dittlinger, whose granddaughter, Maria Leibscher, recalled her grandfather's enthusiasm. "My grandfather was just as enthused as Joe was. Those two fellows, when they went out on their rides, I mean they explored the county, as no one would have explored it!"[25]

During the summer, Hip Mengden often tagged along with his grandfather Dittlinger and Joe Sanders. "He [Joe] was very interested in compiling this accurate map for the county," Mengden remembered. Dittlinger often planned his routine around the mapping excursions. He would go to the office in the morning, work, then have lunch in the afternoon. Dittlinger would rest till 2:30 or 3:00 P.M., then he and Joe would plan the trip. "They knew where they were going and which way they would get back home," Mengden recalled.

On many of these sallies, Joe drove Dittlinger's 1929 Franklin. It was a large seven-seater with plush upholstery, but it could be an ordeal to ride in for young Hip. "If you were bare-legged it was murder! It was prickly ... horsehair. This is hard wear stuff."

And so Joe Sanders worked nonstop on surveying every scenic byway and keeping information for making road signs. Little by little he pieced together each section of Hill Country territory, but the challenging project, much of which still lay ahead, seemed an enormous undertaking for one man.

Chapter Six

Showing You the Way

Joe worked diligently to log all of the backroads. His travels took him along lonely caliche byways, on rocky hillside trails, down muddy paths, and across fords. Sometimes recording routes and distances seemed to be the easy part of the process. Though most ranchers were friendly and cooperative, a few did not see the point of marking a road. After all, local folks had traveled the Hill Country byways for years. Who needed outsiders passing through? These concerns were understandable, but the fact remained that there was an increasing flow of lost Sunday drivers on the backroads.

"Some people just don't like a change," Laurie commented. "There was an attitude of 'what was good enough for Grandpa was good enough for me.'"

But the occasional debate over the roads in the hills was nothing compared to the argument in town over the proposed State Highway No. 2 bypass. Ever since the Texas State Highway Department had announced the construction of a new thoroughfare, controversy arose over whether its course would run through town or along the outskirts. The old highway went right through the middle of New Braunfels, down Seguin Street to the town plaza and west on San Antonio Street. Local merchants wanted to maintain the route and felt that a bypass would hurt business. The heavy traffic, including large trucks and buses, was hard on the roads, however. Joe supported the construction of a new road just outside of town, and sometimes he joined in lively discussions with the businessmen.

Eventually the highway department planned construction for the outskirts of New Braunfels. Comal County agreed to pay for some of the road building and also to finance work on state highways 66 (which followed the present route of U.S. 281) and 46 west of town.[1]

But highway department money and labor did not extend to the backroads, and Joe's chief concerns were with marking them. His exhaustive treks through the Hill Country included a visit to the Kaderli Ranch on the Guadalupe River. Milton recalled, "I remember him [Joe] coming to the ranch and speaking with my Dad and talking about signage and this sort of stuff.... He would come down, I think, every road that there was to find out about...."

As Joe compiled information for a road map and markers, he also researched the specifications for the signboards themselves by examining highway department manuals and signs. An organized distribution of road markers was a relatively new idea even on the state level. The highway department first prepared a statewide marking system in the early 1920s. Not until 1929 did the department post state and federal route signs along all designated highways in Texas. The early guideposts were rather small and displayed road numbers and directions to the nearest city. They were wooden boards at first, and then highway crews erected metal ones but soon returned to wood when the metal rusted badly.[2]

Early highway maintenance manuals directed that "signs should be made of cedar or poplar boards, securely fastened to cedar, heart pine or creosoted pine posts" at least six inches in diameter and set in concrete. The lettering had to be black on a white background. In addition to giving detailed instructions, manuals emphasized the increasing importance of road markers for the benefit of travel.

The construction of hard-surfaced roads changed the character of traffic greatly. On unimproved roads, traffic was largely local or neighborhood. With improved roads, the traffic changed to intercounty, state, and interstate. Tourists were everywhere. The state road system developed rapidly and made it necessary to direct traffic from one section of the state to another. Signs at road crossings and junction points became more and more important.[3]

Once Joe had compiled all the mapping and mileage information for a section of the county, he set out to make road signs. He acquired two-by-two-foot sheets of half-inch plywood from Cameron's lumberyard, a local business on Castell Street. With paint and stencils for the lettering, he began his project. Young Hip Mengden recalled seeing Joe hard at work. "He'd get [the signs] and seal them first. The wood was all sealed, and then white background went on it, and then two colors … they were red and black." Joe printed the various roads, towns, and mileages with black paint, and he painted directional arrows in red.

His project stretched into the spring and summer of 1932 as the work was precise, tedious, and time-consuming. Any spare time he had, evenings or on weekends, he spent on sign making.

"He was obsessed with it—a perfectionist," Laurie observed. "He always said the only way to do a thing was to do it right.… Joe used to sit up all night sometimes working on a sign. [The project] took months."[4] Each sign took about a day to finish. As his work continued, signboards piled up—in the garage, in the living room, and in the kitchen. During weekends he spread everything out on the supper table. On Sundays, with the kitchen under siege, instead of having a homemade supper, he went downtown to the A Cafe and brought home a roast beef dinner.

With the intensity and time the work involved, Laurie admitted that, though she enjoyed logging the roads and viewing the scenery, she became frustrated with the actual sign-making process from time to time. Increasingly, signs took up more space, and she had housework to do.

As Joe completed more markers he faced the prospect of erecting them and realized that he simply could not carry out the job alone. At the local American Legion meeting on July 10, 1932, members elected Joe as Post Commander for 1932-1933. They also decided "that the Post is sponsoring a road marking program at this time and hopes that this move will eventually be a credit to the Post." The executive board moved "that a

Public accounting of the Road marking contributions and expenditures be made."

Laurie remembered, "He was very dedicated to the work in the Legion and thought that would be a good thing to sponsor.... He also thought that it would generate more response to the American Legion as a group than to just one fellow."

The Legion minutes of August 9, 1932, recorded: "Post Commander Sanders reported that the road marking signs are being prepared and work of placing same will be commenced." As Joe continued to paint more signs, however, it took time for the actual posting to begin. During the fall and winter priority went to other Legion activities such as district conventions, an Armistice Day celebration, and a program for making Christmas charity baskets. The Legion held the major responsibility for helping needy veterans hurt by the Depression. Throughout all of this, the fledgling Legion Post had needs of its own and depended on the generosity of local businesses and individuals. There was no Legion building at that time. Members met in various places in town—the courthouse, a dry goods store, or the basement of the A Cafe (where they liked to play poker).[5]

With the spring of 1933, however, came the revival of the project. By this time Joe had enlisted the services of fellow Legionnaire George Eikel-Schmidt in making the signs,[6] but the heavy manual work of posting them still lay ahead. Joe asked for the help of Legion volunteers. Some donated needed tools and materials. A corps consisting of about six men worked with him to erect the markers. They secured sturdy posts to hold the signs. Armed with assorted tools—pick axes, crowbars, and posthole diggers—the men set out. At some intersections it was necessary to break through solid rock to place the posts in precise spots. Workers needed heavy machinery and dynamite. Clarence Rice, who later became county judge, remembered that the Legionnaires borrowed an air compressor from the county in order to dig some of the holes.[7]

Laurie commented, "They had to blast [with dynamite] sometimes for the signs." The volunteers worked whenever they could get time off from

Hardworking Legionnaires use heavy machinery to cut through solid rock to post road markers. Upon completion of the project the signs provided necessary directions to travelers in Comal County. From left: George Eikel-Schmidt, Joe Sanders, A. D. Nuhn, Anton Reininger, Alvin Dauer, Udo Hellmann. (Laurie P. Sanders Collection.)

their jobs during the week and on Saturdays. The labor was difficult, tiring, and hot.

Joe devised a system of numbering the signs, and they erected the first ones near the hamlet of Gruene.[8] Through each number, marked on the back, he kept track of every board's contents and precise location, which he recorded and diagrammed in a logbook. The corps of Legionnaires traveled along the Hancock and Purgatory roads toward the northern part of Comal County, on to the Devil's Backbone area (where two of the signs were posted in Hays County),[9] down the Guadalupe River Drive, and then west on State Highway 46 toward Startz Hill and the backroads beyond. The men started out early in the morning ready to mark a specific Hill Country area, and sometimes the task stretched all day—leaving an exhausted crew.

A brief newspaper note illustrated this point:

> Two Legionnaires were seen looking for three post hole digging bars at the intersection of Curry Creek Road and Mustang Hill Road where the

last road marker was set up about 10 o'clock Friday night. The County Clerk and the Post Commander are still sleepy. Bars still missing.[10]

In spite of the trials and tribulations, the project did not go unnoticed. It even garnered national attention in the March 1933 issue of *The American Legion Monthly*, which published a photo of the workers. "If you've ever been lost forty miles from nowhere, you'll applaud what Comal Post of New Braunfels, Texas, is doing to mark the crossroads in its county."[11]

At the Legion meeting on March 20, 1933, the organization, with a total membership of seventy-three men and $60.60 cash on hand, passed a motion "to authorize the expenditure of approximately $10.00 to complete the road marking program," and "Commander Sanders and Haas offered to secure the posts necessary."[12]

Two months later at the May 26 meeting, minutes show that members were pleased to hear that Legionnaire and local rancher, Arno Knibbe, supplied a sufficient number of cedar posts "to complete [the] road marking program. Gratis."

Legionnaires of Comal Post 179 proudly display their handiwork. The photo appeared in *The American Legion Monthly* for March 1933. From left: Alvin Dauer, Egon Hoeke, Udo Hellmann, Joe Sanders, Fred Pfeuffer, Anton Reininger, Bob Tays. (Laurie P. Sanders Collection.)

During these months of work, Joe also finished his county map. With all the roads logged and hamlets and tourist camps taken into account, he was ready to have the map drafted and printed. He enlisted the help of local resident C. H. Bernstein, who worked as a civil engineer in San Antonio, to draw up the chart. The entire Comal Legion Post 179 received credit for the road map with J. J. Sanders listed as compiler and C. H. Bernstein as draftsman. A number of local businesses sponsored the production of the map: Gerlich Auto Company, Schumann Store, Henne's, Peerless Pharmacy, the A Cafe, Dittlinger Roller Mills and Dittlinger Lime Company, Eiband and Fischer, and two resorts—Slumber Falls Camp and Landa Park.[13]

Newspaper reports came out in anticipation of completion of the Legion project:

American Legion Road Marking Is Under Way

Post Commander Sanders reports that the signs for marking the various intersections of Comal County roads are ready for placing and calls for volunteer Legionnaires to help complete this work. To further aid in this work, it is hoped that some enterprising citizen will offer the use of a truck to carry the signs to their proper places.

The carrying out of this road marking program was made possible only after several grateful citizens made donations to the post for this purpose, namely Mr. Fuchs, Mr. Dittlinger, Mr. Alf. Staats, Mr. R. A. Ludwig, and Miss Lottie Druebert. Paint was donated by Mr. Alvin Plumeyer and a very reasonable charge for labor done by Meckel Bros. and Geo. Eikel-Schmidt.

The various roads were logged and signs planned by the local Post, Legionnaire Sanders, chairman of the committee, who also worked out a country road map, showing the various roads with distances from the Plaza to the various intersections. Quite a number of these maps have been placed with local citizens, and it is hoped that in the future a smaller copy of this map can be gotten.

To all those that in any way assisted in this program, including all those who purchased tickets for the dance held for this purpose and to Judge [Carl] Roeper for his co-operation goes the gratitude of the Legion.

The signs are all complete and will be erected as soon as the volunteers and light truck are available.[14]

Legion Completes County Road Map

A complete road map of Comal County has just been completed by J. J. Sanders, Post Commander, who compiled it for the Comal Legion Post 179.

The map, one of the first of its kind for Comal County, has listed and routed every road in the county, named and located every village and hamlet, and gives mileage between various points.

The route of the Guadalupe River is carefully traced on the map with important points named. All roads leading out of the county to some other town are so designated with mileage given.

The map was drawn by C. H. Bernstein after figures and data had been provided by J. J. Sanders.

Copies are available at the Chamber of Commerce.15

According to Legion minutes, by the summer of 1933 the men had completed marking all the scenic Hill Country roads in Comal County:

A road-marking program was worked out by Commander J .J. Sanders and aided by Adjutant R. H. Tays, Tony Reininger, A. D. Nuhn, Geo. Ei[ck]el-Schmidt, Alvin Dauer, Fred Pfeuffer, U. R. Hellmann, Richard Ludwig, Egon Hoeke, Harry Wehe and Past Commander Haas is being sponsored and financed by Comal Post No. 179. The money for this undertaking was raised through the medium of a dance. Every county road was marked for distance, sharp curves, dips, guards, narrow bridges, river water-gauges, etc. The work required 81 signs and about one-half of the postholes had to be drilled and blasted. In addition to this work, a scenic road map was compiled by Sanders for the Post with a free distribution of over 5,000 copies.[16]

With the summer touring season in full force, there was quite a demand for the county maps that, along with the road markers, made the countryside more accessible to scenery lovers. Different organizations interested in exploring the Hill Country contacted Joe and inquired about his map. From their office in Houston, the Motor League of South Texas, a division of the national American Automobile Association (AAA), wrote Joe for tourist information on July 22, 1933.

Dear Sir:

We are anxious to secure some reliable information concerning resorts, vacation places, playgrounds, etc, and check up on the character and conditions of the roads leading into those places located in your section principally in that part of the country west of you within the territory bordered by State Highways 9 and 2 south of No. 20. The San Antonio Auto League, our San Antonio office advises we write you for some copies of an interesting map of Comal County.

Will you kindly send us two of these maps, mark out on one of them for our records, the roads that lead to any point that would interest folks visiting for health, recreation or just pleasure; pointing out character of roads such as if paved, if graveled or if dirt. Give me the name or names of dude ranches, camping grounds, etc. Kindly write such places and ask them to write me giving me in full detail what attractions they have to offer, what accom[m]odations, what prices. We are continually called upon to find places for hundreds of folks locally and all over the country throughout each season and we want to take care of these inquiries. This service we render [is] very valuable to all parties concerned except possibly ourselves as there is no reward except service rendered our state. I know the above details quite a bit of effort but on a recent survey of your

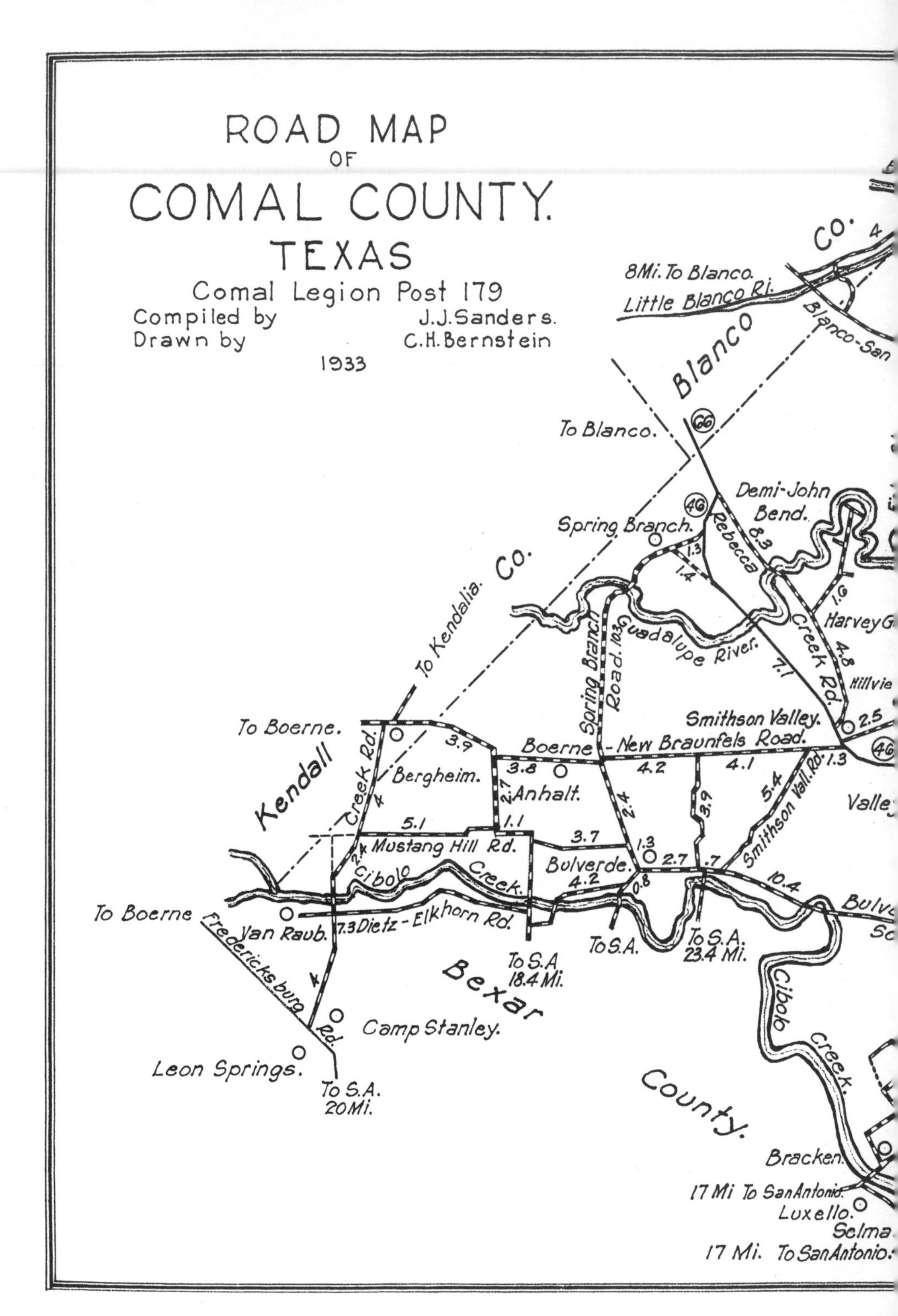
ROAD MAP
OF
COMAL COUNTY.
TEXAS
Comal Legion Post 179
Compiled by J.J.Sanders.
Drawn by C.H.Bernstein
1933
8Mi. To Blanco.
Little Blanco Ri.
Blanco Co.
To Blanco.
66
46
Demi-John Bend.
Spring Branch.
Rebecca
Creek Rd.
Guadalupe River.
Spring Branch Road. 103
To Kendalia.
Kendall Co.
To Boerne.
Smithson Valley.
Boerne - New Braunfels Road.
Bergheim.
Anhalt.
Creek Rd.
Mustang Hill Rd.
Cibolo Creek.
Bulverde.
Smithson Vall. Rd.
To Boerne
Fredericksburg Rd.
Van Raub.
7.3 Dietz - Elkhorn Rd.
To S.A 18.4 Mi.
To S.A.
To S.A. 23.4 Mi.
Bexar County.
Camp Stanley.
Leon Springs.
To S.A. 20 Mi.
Cibolo Creek.
Bracken.
17 Mi To San Antonio.
Luxello.
Selma.
17 Mi. To San Antonio.

Road Map of Comal County...
1933

> section of the state I found out how little is known of it and it certainly needs being brought to the attention of the tourist.[17]

San Antonio papers touted the allure of the magnificent scenery in the hills of Comal County. Writer and New Braunfels resident Gertrude Rawson, in a feature about a Hill Country tour, wrote:

> ... motoring in Comal County has been facilitated by an excellent map made by J. J. Sanders for free distribution, and indicating all the places where the view is especially fine. Every corner and cross road in the county has been fully and plainly supplied with signs, an enterprise of the Comal Post 179 of the American Legion.[18]

In recalling his idea and service in the Legion, Joe stated, "It was a real pleasure putting every spare hour I could find into the work." Comal Post 179, with all the hard work, was well on its way to claiming the slogan: "The Legion Showing You the Way."[19]

Chapter Seven

Celebration and Controversy: The Road to the Centennial

In the summer of 1933, after posting almost all of the Comal County roads in the Hill Country and along the Guadalupe River, Joe hoped that the lost motorist would be a character of the past. The Sanders family often went on drives to admire the scenery and Joe's handiwork. But it did not take long, only weeks after Legionnaires struggled to erect the signs, in fact, to find some markers torn down. The family suspected that some of the placards were taken down by people who didn't want to encourage traffic in the area or through their ranches. "But mostly it was people who, I don't really know for what reason, [were] just destroying them," Laurie Sanders commented. They found signs knocked to the ground or thrown off to the side. "They'd run into them with the car or truck," she added.

Vandalism, though an uncommon occurrence, was not new. As early as 1898 Comal County commissioners issued a $100 reward for information to convict the person who tore up "the 8, 10, and 13 milestone[s] on the New Braunfels and Smithson's Valley road."[1] And so Joe's scenic drives soon turned into routine excursions to inspect the condition of the markers as he took it upon himself to replace damaged or weathered signs.

"We were always driving up there to see how they were, and if he saw one that needed [repairs], well he'd take it down and bring it in and fix it up right away, [and] take it back up there," Laurie recalled.

Occasionally Joe went on sign-posting trips with Mr. Dittlinger. Grandson Hip Mengden remembered the freshly painted boards loaded up and ready to replace battered ones. "If there would be some signs to put up, well then the signs were already painted and in the back of the car." For stakes that were completely toppled, the nearest fence post had to suffice to hold a signboard. Joe was determined to replace each marker if necessary and keep the backroads friendly for tourists.

The promotion of safe and scenic motoring was beginning to take hold on the state level. Transportation and tourism figured prominently in the growing Centennial movement in the early 1930s, as Texans prepared to commemorate 100 years of Texas independence. As Joe's term as post commander drew to a close he appointed a Centennial Committee composed of himself and six other Legionnaires to represent Comal County in "carrying forward the program of the Texas Department of the American Legion to 'Make Texas Centennial Conscious.'"[2] They circulated rosters to enlist membership in a Texas Million Centennial Club to increase public awareness and support for a statewide celebration in 1936.

In 1933, the state highway department created the office of landscape architect as part of a campaign to improve the safety and aesthetics of public roads. The department also appointed Mrs. Frank W. Sorell of San Antonio as head of a Texas citizens' highway beautification group to foster public support and involvement. Under her leadership local road clubs promoted highway improvement in their areas.[3]

In New Braunfels the push for road construction and advancements, however, was becoming mired in a growing controversy. The town's main artery, Highway No. 2, was once again the focus of concern. Ever since the Texas Highway Department announced plans to reconstruct the road, Comal County officials had been in constant contact with state officials, who proposed a new route on the outskirts of New Braunfels with a Y-extension going into town. The department planned construction of a Guadalupe River bridge on the new thoroughfare, as well as paving the entire highway. Comal County's portion of Highway No. 2 would cost an estimated $2,300,000. To aid in the project county commissioners pledged $25,000 to acquire right-of-way along the course, but by late summer 1933 the county had already spent $41,000, and the city of New Braunfels had spent an additional $11,000.[4]

Comal County officials were clearly upset with spending excessive funds, and, as the local newspaper reported, "even diverting County Road and Bridge money, sorely needed on Lateral County Roads to highway construction." County commissioners sent a memo to the State Highway Commission outlining their concerns. A major problem was the "dust menace" caused by the breakdown of highway materials under the strain of heavy traffic and by a lengthy drought. The commissioners also complained of sudden difficult turns to property off the highway, the lack of culverts, diversion of water flow, and poor maintenance. In their frustration they pointed out that the road project had been dormant for over six months due to an apparent lack of state funds; the state, however, had appropriated money for other Texas highways.

The situation had driven "residents and business people adjacent to said highway to the breaking point." The commissioners court implored the State Highway Commission for action:

> Even though the citizenship of our county is of such a class as will endure under the most trying and adverse circumstances, shirking no duty imposed by any department of Government ... we would ask the Department an equal measure of compliance on their part, even as we have complied regarding that, of which we have submitted our prayer, and we think and feel that we are justly and rightfully entitled.[5]

Another road scheduled for improvement was Texas Highway No. 46. Engineers planned to widen the road to include a 100-foot-right-of-way, straighten perilous curves, add topping, and extend the route through New Braunfels and toward Seguin to the Guadalupe County line. Controversy regarding the work on Highways No. 2 and 46 even led New Braunfels and the local Good Roads Committee to refuse right-of-way for improvements to State Highway No. 66 going through the northern and western sections of the county until the completion of the other two roads. Committee members voiced their opinions to State Highway Commissioner D. K. Martin at a luncheon in San Antonio. Their defiance evidently hastened the commissioner to promise to expedite the projects.[6]

If Comal County felt that state authorities were dragging their feet, local officials initiated what county projects they could by funding several small road repairs. The federal government provided some relief regarding local road work during the difficult Depression years. First, the Reconstruction Finance Corporation under President Herbert Hoover furnished funds for part-time jobs until the late summer of 1933. Franklin Roosevelt's Civil Works Administration and later the Works Progress Administration were among several New Deal programs that provided money and manpower. In Comal County a relief committee administered various employment programs that aided the execution of local building projects, including road work. In late 1933 road crews topped several byways with gravel or caliche, including portions of Koepp Road in southern Comal County, Bulverde Road to the west, Watson School Road off Highway No. 2, and York's Creek and Meyer roads in the eastern part of the county. The city also provided for the grading of some municipal streets.[7]

Debate and discussion over road projects and controversies were so prevalent that by the end of 1933 the editor of the *New Braunfels Herald* declared his number one New Year's resolution to be "No more Post Road Headlines."[8] In 1934, however, the persistence of Comal County citizens and officials began to pay off, at least in part, with the start of work on Highway No. 2. To "celebrate the laying of the first pavement on the new San Antonio-Austin Post Road," the city declared May 10, 1934, as Highway Day. New Braunfels threw an "old-fashioned barbecue" to "show appreciation of the better travel facilities afforded them," and invited state highway commissioners and representatives of area cities.[9]

The anticipated special event even led New Braunfels mayor H. F. Fischer to designate Highway Day as "Prosperity Day" also and to "call upon all citizens, public and private to resolve to celebrate such day as marking a New Era for the city from which to march definitely forward and upward into a new day and a new deal for community and our section."[10]

Over a thousand spectators witnessed a "mile long" parade that ended at Landa Park "where barbecue and beer was served." Numerous civic groups participated, and declared the affair an overwhelming success.[11]

The beginning of this great new highway prompted renewed enthusiasm for promoting easier and friendlier travel, and a number of organizations in New Braunfels became involved. After Joe Sanders' tenure as Legion post commander, the local chapter maintained a Highway Marking Committee. Joe served as chairman. Comal post's services over the previous year had been recognized by the Texas Department of the American Legion who awarded the Community Service Citation to Comal Post 179 for "outstanding service in this community during the past year."[12] On May 25, 1934, the *New Braunfels Herald* announced the completion of the final stage of road signs in the southern part of the county. They were posted on the best "all-weather road from New Braunfels to Marion (in Guadalupe County) by J. J. Sanders." With the virtual completion of the project, the *Herald* concluded, "a driver can hardly become confused."

Besides the Legion, other civic organizations, including the Junior Chamber of Commerce and Board of City Development, participated in road awareness projects. The most active group working for highway improvement was the Comal County Highway Beautification Committee. Under the leadership of Patty Fuchs, the committee, a subgroup of the state organization, sought to promote tourism and travel through landscaping and the creation of roadside parks. Her husband, John R. Fuchs, later wrote, "This appealed to her aesthetic nature."[13]

Roadside beautification provided more benefits than just aesthetics. The Texas Highway Department proposed that "beauty could be practical," and the planning and construction of a highway included four principal objectives: safety, convenience, comfort, and aesthetic value. Engineers endeavored to shorten routes, eliminate dangerous curves, smooth out deep culverts, and broaden road shoulders. Planting flowers, trees, and shrubs not only enhanced the scenery, it also prevented erosion along roads and reduced the need for highway maintenance.[14] Roadside parks and turnouts were beginning to be considered necessities of convenience and safety that would "pay dividends of pleasure" to travelers, as *Texas Parade* magazine descriptively emphasized:

> If you are among that great class of Americans characterized as "the average motorist," you can well remember driving along a highway on a torrid summer's day. You have come many miles since morning. You are tired. The merciless sun is getting on your nerves. Even the efforts of alert filling station attendants, who have done everything but wash behind Junior's ears, leave you yearning for an unnamed, almost intangible something to give you surcease from the driving fatigue that sometimes turns motoring into an arduous and unpleasant task. The kids, if you have them along, are hot and restless, and as you spin along, you cast envious glances at a grove of trees bordering the right-of-way, but just inside the farmer's fence.
>
> Wouldn't it be elegant if you could get over there, roll in the lush grass and let your temperature drop in the comforting bosom of that gloomy and entrancing shade? But that high fence keeps you out and, unconsciously, you press the accelerator a bit harder.[15]

And so the state set a goal to establish 1,000 "off-the-road nooks" by the end of 1936 to "offer rest and relaxation after a ride in a broiling sun." During tough Depression times funds weren't readily available, however. In order to make "Texas highways the most beautiful in the United States" people had to make up for the lack of money with "imagination, hard work, the art of scrounging, and the use of materials at hand."[16] With this in mind the highway department and local groups selected natural beauty spots along highways—incorporating existing trees and shrubbery. Generous citizens donated supplies and labor to set up picnic tables and plant more vegetation. Federal assistance from the National Youth Administration, created by the Roosevelt administration in 1935, also provided some help around the state, employing young men to build parks.[17]

Women's groups were probably the most instrumental in "dressing up Texas landscapes for the Centennial." Through educational programs they encouraged farmers to clear their roadside property of high weeds—aiding the motorist's view of curves in the road ahead. They also urged gas stations and tourist courts to keep their businesses clean for travelers. They educated the public about laws against disturbing natural vegetation and pulling flowers, and they lobbied to restrict billboard advertising and livestock grazing along state highways.[18]

In Comal County the local beautification committee held an election to choose an official tree and shrub. The mountain laurel won as best shrub with 100 votes, and the Spanish oak and live oak received seventy-six votes each to serve as the official trees. Groups planted these selections along roadside parks and at the entrances to town. They also incorporated other native vegetation such as ceniza, crepe myrtle, mountain cedar, agarita, and lantana.

The local committee set out to acquire beauty spots, especially along Highway No. 2. The "reposes" were to be "within sight of the highway" where travelers could "stop for rest and recreation."[19]

Laurie Sanders remembered, "Usually they'd find a spot where there might already be a little scrubby tree or something, and they would put out a table where you could sit...." The first official roadside park in Comal County would be established on State Highway 66 on four and one-half acres of land donated by Hermann Knibbe, Sr. The committee awarded road beautification prizes to farms and ranches with the most attractive roadside appearances and to rural schools and filling stations for tidiness.[20]

To commemorate the upcoming Centennial celebration, roadside beautification committees also promoted historical sites across Texas. Mrs. H. Dittlinger, as a member of the Texas Centennial Commission, played a key role in helping to establish historical markers in Comal County and the state, and Joe chauffeured her to dedication ceremonies all over South Texas. Hip Mengden commented, "They would leave, see that was before the age of air conditioning, so they'd normally leave at four o'clock in the morning to get to where they were going and back before it got hot—really hot. Might be 100 miles each way."

In the couple of years before the Centennial, Joe was chairman of the road-marking subcommittee of the county highway beautification group, and he attended local- and state-level meetings as necessary. He continued to maintain county signs and erect new ones on the way to scenic spots such as the Narrows. But the posting proved to be a difficult and ongoing job because vandalism had gotten worse.

"A lot of [the signs] had been shot at, you know, target practice I guess," Laurie surmised. Her daughter suspected that the perpetrators were "just a bunch of toughs going out at night probably."[21] At any rate the damage finally prompted the *New Braunfels Herald* to print a plea to end the problem.

Legion Asks Regard For Co. Road Signs

Reports that road signs marking scenic drives through Comal County have been changed or mutilated have caused the American Legion road-sign committee, which is responsible for their erection, to ask that the public assist them in maintaining the markers.

J. J. Sanders, chairman of the committee, has requested that persons wishing additions or changes made in the information on the signs should communicate with him in order that arrangements may be made to do the work neatly.

Sanders asked that motorists inform him of mutilated or damaged signs, so that they may be repaired promptly. The Legion is now at work placing the markers in perfect condition preparatory to turning them over to the county government for maintenance.[22]

As Joe worked on the upkeep of the signs, he also corrected and updated information on the road map. He wanted to produce a new version in honor of the Centennial. The family still spent their excursions exploring new byways, except that now Joe traveled the roads in another vehicle—a large six-passenger Franklin he had acquired from Mr. Dittlinger. The early 1930s model Franklin was equipped with a double transmission that gave it more traction and power. Another passenger, a wire-haired terrier named Billy Boy, was a good companion for nine-year-old Laurie Jo. He was an eager rider who poked his head out the window to sniff the country air.[23]

Joe intended to make the new map more interesting and eye-catching to tourists. He enlisted the help of Phil Rawson, an artist and architect who practiced his own style of artistic engineering and referred to his work as "Imagineering."[24] His wife, Gertrude Rawson, wrote tourism articles for the *San Antonio Express*, and the Rawsons were interested in Comal County's natural wonders. As Joe redid his map, P. N. Rawson drew small sketches on it, locating each scenic attraction, as well as inserting wild animals such as deer, foxes, wildcats, and armadillos. The map depicted roads in solid black, broken, or lightly drawn lines indicating the condition of various byways—paved, all-weather, and dirt.[25]

Joe informed the Legion of his plans at a meeting on January 6, 1936. The minutes read, "that in case Comrade Sanders needed financial aid in completing the road map, that the Post assist him to the extent of $50.00, if necessary."

He produced small black and white editions of the map, as well as large copies (twenty by twenty-four inches) printed in color. Both the Chamber of Commerce and the Legion distributed them. Community and area response was enthusiastic, and as Joe sent promotional copies to state officials, he even received acknowledgment and thanks from Governor James V. Allred, who wrote on February 29, 1936:

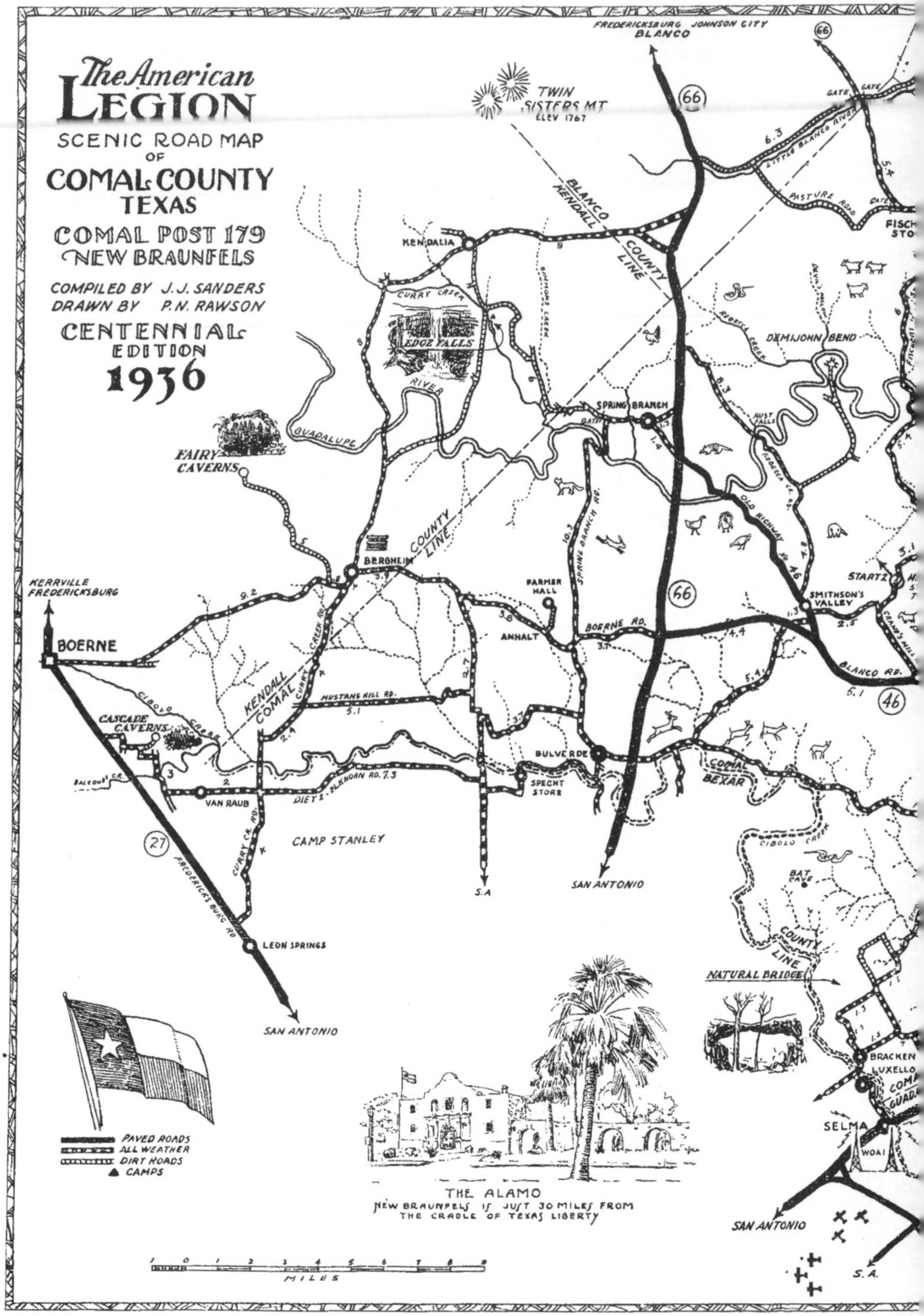

The American
LEGION
SCENIC ROAD MAP
OF
COMAL COUNTY
TEXAS
COMAL POST 179
NEW BRAUNFELS
COMPILED BY J. J. SANDERS
DRAWN BY P. N. RAWSON
CENTENNIAL
EDITION
1936
FREDERICKSBURG JOHNSON CITY
BLANCO
TWIN SISTERS MT
ELEV 1767
BLANCO
KENDALL
COUNTY
LINE
KENDALIA
CURRY CREEK
EDGE FALLS
RIVER
GUADALUPE
SPRING BRANCH
DEMIJOHN BEND
LITTLE BLANCO RIVER
PASTURE ROAD
FAIRY CAVERNS
BERGHEIM
COUNTY LINE
OLD HIGHWAY
SPRING BRANCH RD.
FARMER HALL
SMITHSON'S VALLEY
STARTZ
KERRVILLE
FREDERICKSBURG
BOERNE
ANHALT
BOERNE RD.
BLANCO RD.
KENDALL
COMAL
MUSTANG HILL RD.
CIBOLO CREEK
CASCADE CAVERNS
BULVERDE
COMAL
BEXAR
VAN RAUB
SPECHT STORE
CAMP STANLEY
CURRY CR. RD.
FREDERICKSBURG RD
S.A
SAN ANTONIO
CIBOLO CREEK
BAT CAVE
COUNTY LINE
LEON SPRINGS
NATURAL BRIDGE
BRACKEN
LUXELLO
SELMA
WOAI
PAVED ROADS
ALL WEATHER
DIRT ROADS
CAMPS
THE ALAMO
NEW BRAUNFELS IS JUST 30 MILES FROM
THE CRADLE OF TEXAS LIBERTY
MILES

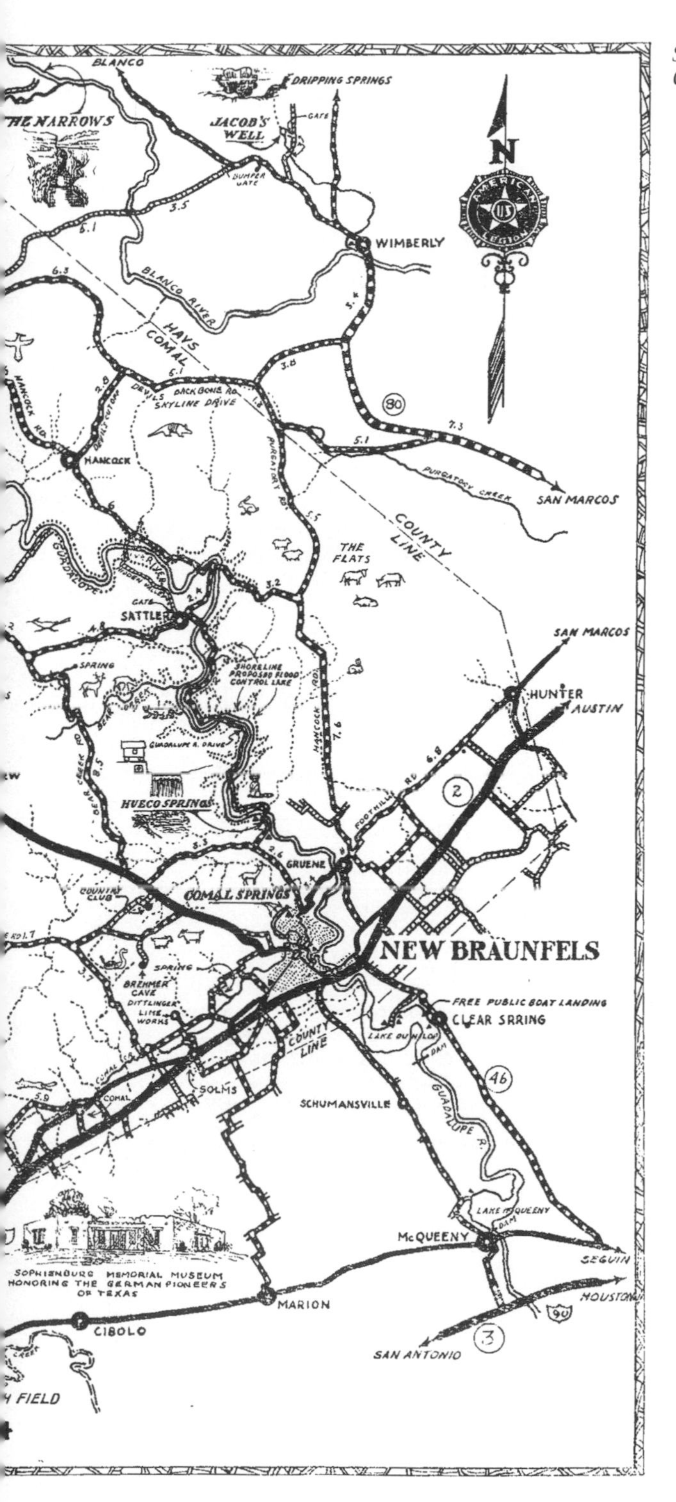

Scenic Road Map of Comal County...Centennial Edition, 1936

My Dear Friend:

I want you to know how deeply appreciative I am for your thoughtfulness in sending me the Centennial map of your American Legion post of Comal County.

This is indeed a most interesting and attractive scenic map, and I want to wish you every success in the sale of these maps for your post.

With kindest regards and best wishes,

I am, Sincerely yours,
James V. Allred
Governor of Texas

The Centennial Edition encouraged tourism in Comal County and promoted its many attractions to Centennial travelers. This was at a time when 1,525,579 registered vehicles in Texas traveled over a state highway system with a total of more than 21,000 miles.[26] As families, including the Sanders family, prepared to journey to the Centennial Exposition in Dallas, numerous motorists drove along Highway No. 2 and stopped off in New Braunfels and Comal County. Area newspapers, including the April 23, 1936, edition of the *Blanco County News,* advertised the map.

This handsome map of Comal County and the Hill Country adjacent to it in Blanco, Hays, and Bexar Counties, was prepared by the American Legion of New Braunfels, and shows every road and place of interest in the entire region. It bears out the statement made last week by A. W. Grant, managing editor of the San Antonio Express, that Comal County has done more than any other county in Southwest Texas to mark and give publicity to its roads and beauty spots.

Among the Hill Country sightseers was A. W. Grant himself, who easily negotiated the backroads. It was evident that stopping to ask Joe for directions years earlier had paid off, and after planting the seed for Joe's idea, Grant became his most ardent supporter along the way.

. . . . Part 2 . . . .

HILL COUNTRY DRIVES

"... look about you."

Chapter Eight

The Guadalupe River Drive

Hill Country byways presented endless opportunities for nature lovers to explore out-of-the-way scenic wonders. The 1936 Comal County map guided motorists, while the reverse side offered helpful suggestions, statistics, and descriptions of the points of interest:

> As you drive along Highway No. 2 between Austin and San Antonio you may notice to the northwest a low line of cedar covered hills. This is the Balcones Escarpment along which, ages ago, the earth split; and the coastal plain on which you are traveling, fell away several hundred feet.... Time has softened the outlines of the escarpment so that at places it is barely discernible. But behind this ancient rampart lies a scenic wonderland, known as the Texas Hill Country, so varied and unexpected in its charm that it will reward an afternoon's fast travel over its skyline drives or weeks of exploration in secret places where dark waters reflect the feathery moss-draped forms of giant cypress trees.
>
> You are handed this complimentary map of Comal County in the hope that it will fire your imagination and make you pause, if only for an hour, to look about you. We believe you will find yourself rewarded by scenes that will never leave your memory and which may even be the reason for your returning to New Braunfels time and again.

The advice on the map also included a reassuring note that all the road intersections in the county were plainly marked. "This project was sponsored by the American Legion for your convenience. You need have no fear of losing your way out there."

And so, by the mid-1930s the Sunday driver was well-equipped to tour the Comal County backroads and search for some breathtaking vista just around the next curve. One newspaper feature described the countryside northwest of New Braunfels as "a land of high blue hills and shady river valleys and winding roads alluring to the motorist."[1] Probably the best known and loved scenic road was the Guadalupe River Drive. The map read:

> [T]his road gives you an opportunity for first hand acquaintance with the fascinating and temperamental Guadalupe River. In places the right of way includes the shoreline. At other points you have access to privately owned picnic grounds. This narrow strip of country presents endless recreational possibilities.

The Guadalupe River Drive began as a small country lane laid out for the benefit of farmers and ranchers living in and around the valley. In 1900 landowners Peter Nowotny, Jr., Eduard Kanz and others signed the original road petition with supporting motions made by commissioners August Schulze, Jr., and Ben R. Smithson. A survey was conducted from November 1900 to February 1901, and an appointed jury viewed the layout of the byway the following April. On August 13, 1901, the Comal County Commissioners Court approved and established "a public road, third class, twenty-two feet wide with gates."[2]

The road, which was approximately nine and three-fifths miles long, began "at a gate on Harry Landa's ranch ... on the Austin-Boerne Road" about four miles from New Braunfels and ended "on the Mountain Valley Road on Henry Erxleben's place." The route was little more than a rocky trail cut out of the hillside. The four river crossings were passable only in dry weather. As the age of the automobile dawned, commissioners tried to upgrade the river fords. In November 1917, they called attention to needed

In the early days, the river road was little more than a narrow country trail. Drivers, contending with hanging branches and fallen boulders, often had to get out and clear the path. In many places the road was only wide enough for one car (in this case a Model A). (Photo by Joe Sanders.)

improvements for the crossings. In March 1918 commissioners allocated $475 for a low-water bridge at the third crossing, also known as Spiess Bottom. Concrete contractors Joseph and Albert Meckel did the work, which first included cutting down an embankment. This literal chiseling out of the mountain required hours of labor with jackhammers, picks, and sledgehammers.[3]

Some sections of the road were so narrow that tree branches and rocks obstructed the way. A driver sometimes had to stop and clear a path. When two cars met, one motorist had to back up until the road widened. In view of these difficulties, in March 1925 property owners petitioned county commissioners to change the river drive from a third-class to a second-class road "with a width of not less than 30 feet." In the following decade civic groups asked for more improvements to bring certain rough stretches "up to second-class standard," but they also emphasized that scenic drives should be "available to the community for their charm and isolation rather than for their usefulness as arteries of commerce."[4]

Though the route could be troublesome, its very character was formed by the rugged beauty of its surroundings. As early as the German pioneer days, traveler Heinrich Ostermeyer wrote about the region of the "splendid sparkling Guadalupe" in his journal. He described a land with "towering cypresses and pecans … springs of marvelously sweet water" and an "abundance of game." He concluded that "this Texas river valley [seemed] more wild and romantic than anything the older country had to offer."[5]

When Joe Sanders came to New Braunfels one of his earliest discoveries was the beauty of the Guadalupe River Drive. For him it became a gateway to the quaint hamlets and scenic vistas of the Hill Country beyond. He took Laurie on the drive when she first visited the area, and through the years the river road became a favorite Sunday excursion. They forded the crossings and followed the meandering byway lined with stately cypress trees. Sometimes they spied mountain goats standing on what looked like sheer precipices on the cliffs across the river. At the road's edge were big boulders that had once been part of the hillside. Along the trail springs bubbled out of the cliff sides. "There'd be springs oozing out and running across the road, keeping it kind of slick and muddy in spots," Laurie remembered.[6]

Apparently one spring in particular had been well known to river valley settlers. Melba Meckel Roth, daughter of contractor Albert Meckel, remembered a place called the "Gossip Spring" between the third and fourth crossings on the river road. It was a resting spot where travelers stopped and chatted. "You stopped to get a cool drink of water, and often someone else was stopped there, and so you just visited a while. You were never in any tremendous hurry."[7]

The Guadalupe River Drive was an ambling, leisurely sort of route. One took time to appreciate the scenery. Moreover, a driver had to be prepared to make stops if necessary to clear away branches and debris that washed through after a storm. It wasn't unusual to slow down for cattle on the road either, and there were cattle guards along the way, but these things added to the charm of the excursion.

Dorothy Kerbow recalled many pleasant family trips. "We went on that road a lot. My mother would say 'Let's go to Wimberley by the river road.'"

Hip Mengden also traveled the road. "That whole drive up the Guadalupe ... was a gorgeous drive, 'cause you had all those tall cypresses down there, and the road would follow the river, and it was shady. Even in the summertime, there was a breeze."

Occasionally the Sanders family would stop along the road where there was space enough to park and have a picnic lunch in the car. One memorable picnic actually occurred in the riverbed itself after a previous flood caused a shift in the stream's course and left an islet in the water. Joe found a safe spot to ford and was able to park the car on the isle, and the whole family, including Laurie's parents and her little niece Garnet (or "Wiggie" as they nicknamed her), piled out of the car and picnicked there. They built a small fire and roasted ears of corn, passed around sandwiches and chips,

A view of the Guadalupe River Drive (ca. 1932) just past the first crossing. The distant figures of Laurie and Laurie Jo can be seen walking on the road. Motorists were free to stop and explore. This photograph was later made into a postcard. (Photo by Joe Sanders.)

The Pfau and Sanders families enjoy a peaceful picnic at the Guadalupe River about 1930. (Photo by Joe Sanders.)

and brought an ice chest full of cold drinks. Trees growing on the island provided shade, and trickling all around was shallow cooling water. The sandy bottom glistened in the sunlight, and little Laurie Jo and Wiggie delighted in the river. The place made for a relaxing, enjoyable afternoon.[8]

Though the area along the drive largely consisted of rugged wild country and old family ranches, a few camphouses here and there served as summer homes, and a few tourist camps went into business.

One early resort that had opened around 1930 was Waco Springs Park. It was located at the first river crossing on the ranch owned by Bob Gode, who leased the park area to Phil Rawson. The Waco (or Hueco) Springs, probably named for the Waco Indians,[9] were just off the road. A small dam had been built, and the water powered a hydroelectric generator. Rawson and his wife Gertrude ran the park, which had picnic grounds, cabins, and offered refreshments and opportunities for boating and swimming.

Gertrude compared the modern park to the wild country of the past century in a *San Antonio Express* feature.

> Instead of tepees, there are pretty little green cottages; where the campfires used to burn are picnic tables and stone fireplaces; and instead of venison, armadillo and jackrabbit, the modern visitor may consume roast duck sandwiches, ice cream and cold drinks.[10]

Park brochures advertised cottages "equipped with springs and mattresses, ice box, electric lights and outdoor fire place" at a rate of $1.25 a day for two people. Bed linen was an additional 25 cents. Renting boats cost from 25 cents per hour up to $1.75 a day. The boats were "light and graceful, easy to handle and perfectly safe," ideal for a "large expanse of quiet water particularly alluring on moonlight nights." For swimmers the resort had dressing rooms and even rented bathing suits for 25 cents. Picnic grounds cost 25 cents per car. Wayfarers could stop for refreshments—a cool drink and some "certain deservedly famous Waco Springs Specials."[11]

One of these "deservedly famous" specials was the roast duck sandwich—a particular favorite of little Laurie and her cousin Wiggie. The Sanders family had to stop and try the special after the two little girls had talked about the sandwiches forever.

While the route offered picnickers a pleasant afternoon and a snack, it didn't serve just Sunday motorists and adventurers. For many area farmers and ranchers it was the major route from the country hamlets into New Braunfels. Writer Gertrude Rawson described the role of the road and its connection to the Guadalupe Valley way of life:

> It serves as a highway for the rather prosperous mountain farms, tucked away snugly in their fertile valleys, and populated almost entirely by Germans whose industry and thrift are indicated by their white-and-blue painted houses and their well fenced pens and gardens near by. The tiny hamlets of Sattler, Fischer's Store, Hancock, Crane's Mill and

Little Laurie Jo (lower right) clings to her cousin Wiggie and seems dismayed that the family picnic is almost over. Laurie's parents, Oscar and Dorcas Pfau, and Laurie Pfau Sanders are in the back left to right. (Photo by Joe Sanders.)

> Smithson's Valley are the gathering places for these farmers; each one has a country store, a filling station, a gin, a school house to which the children ride on donkey back, a bowling alley, and a dance hall which is simply a bare galvanized iron oven during the long summer days, but which blossoms forth one night in the week with electric lights and a hot dog stand and a local orchestra and all the cars for many miles around.... Old and young join in the dancing, and the polka and schottische and "put your little foot" are as popular as the waltz and the one-step. Some of the mountain boys and girls plow or chop cotton all day, dance all night, and then go back to their work on the farm. A simple cordiality prevails, and visitors to these Guadalupe Valley affairs may discover a genuinely delightful rustic atmosphere.[12]

One such "mountain farm" was the Kaderli Ranch on the Guadalupe River. The Kaderli family came from Switzerland in 1845, and the beauty of the Texas Hill Country reminded them of home. They settled on the Guadalupe at a place called the Demijohn Bend. Located upstream from the river drive, this was rugged ranch country, and the unique twists and turns of the waterway formed the outline of a narrow-necked jug—a demijohn.

The scenic road map described it: "Here the Guadalupe River doubles back so close to itself that you can stand on the high ridge, commanding a fine view to the northwest, and throw a stone into the river on either side (If you're good)."

The river was an important part of life for young Milton. He attended the Rebecca Creek School, located over at the north end of their ranch. Each morning he started off early to make the two-and-a-half mile ride on horseback. He would ford the Guadalupe River and trot across the hills to school. At four o'clock, he rode back home ready to start his chores.

"Ever so often ... it would rain hard while we were at school," Milton remembered. "We could cross [the river] in the mornings and in the evening when we'd got back, the river would rise and we couldn't cross." Stranded, Milton had to climb a nearby cliff and call out to his house about 150 yards from the opposite shore. "We'd holler as loud as we could and somebody would hear us." His rescuers rowed a boat across, bringing feed along for the horse. "Next morning we'd cross again by boat, climb up the hill where the horse was and saddle it up and ride to school," Milton said. "There was no excuse for not going to school regardless of what kind of weather it was."[13]

Another rancher, Willard D. Hill, owned the other half of the Demijohn Bend. During Joe Sanders's travels and especially during the course of his mapping the roads and putting up the road signs, he visited both ranches. Laurie remembered that they drove "onto the road where there wasn't much of a road" to reach Demijohn Bend. "It was quite pretty, an unusual place," she observed, "and the thing that was interesting about [it] was that the river was practically tied in a knot there. You see it on both sides, going different directions."

While Joe visited with Willard Hill, Laurie and their daughter explored the area and found interesting geodes and other rocks. "On one place we'd go up where there was a cliff. You could walk along and see all kinds of things embedded in the cliffs—shells and fossils of a fish and all that sort of thing—very interesting."[14]

The fossilized shells and sea creatures were the remnants of the geological history that told of the formation of the Hill Country. Water was the sculptor of the landscape, which was once covered by a shallow sea. All these years later water was still a guiding force. The temperamental Guadalupe River, calm and playful in fair weather, could be violent and destructive after heavy rains. Before the era of damming and flood control the river valley was at the mercy of surging waters brought on by torrential downpours sometimes many, many miles upstream. Travelers along the river drive had to be aware of the current weather situation as well as storms that had occurred upriver.[15]

Upstream on the Demijohn Bend, Milton Kaderli witnessed the river's sudden changes. When the family heard about heavy rains upstream, they ran down to the river and watched the water come in. "I remember the water was about four or five feet, and it had a rolling effect," he recalled. All kinds of debris churned ahead as the surge rolled over the rest of the stream. "If you would ever get caught in something like that, it would be terrible."

Along the river drive, flooding could be especially dangerous. The Sanders family was once almost caught in swollen waters caused by a cloudburst upriver.[16] In July of about 1930 the family had been on a picnic. Again Laurie's parents and Wiggie were along. As they neared the second river crossing, they were stopped by warnings of an approaching wall of water. Laurie remembered the scary incident. "We were driving up the river and we met a fellow coming from the opposite direction having a fit." The man exclaimed, "The water is coming! It's that high! It's coming down the river. Turn around and go back!" So the family quickly turned back and motored across the river at Waco Springs crossing.

Others had gathered near the first crossing—picnickers and travelers who had made their way back to safety. The two little girls Laurie Jo and Wiggie were excited about the possibility of witnessing a deluge and squealed with impatience once they were a safe distance across the river.

"After we got across there was another car slightly behind us—an old dilapidated sedan full of people," Laurie commented. Several small children bounced up and down in the vehicle, and the driver looked like an excited character who didn't know what he was doing. As the car approached the ford, Joe warned, "He's going in the wrong place." A number of others motioned to the driver, waving their arms and yelling, "No, here! Farther down, farther down!" Laurie continued:

> He got into deep water, looked like it might be going into the car almost. 'Course cars were built higher than they are now. He was coming across there. If you knew just where to cross, there was flat rock underneath. It was a little bumpy, but it was more or less flat. But if you didn't, you could get off there. And he was coming and that car was leaning way over. I just stood there, and I knew it was going to flop over.

Knowing that the flood waters were coming, a group of men including Joe waded out in the ford to try to steady the car and push it over into the right track. The panicking driver was desperate for help. "They had to just practically pull him out," Laurie exclaimed. "He was excited, and he couldn't do anything. Of course the kids in there were screaming. He got through all right."

The men guided the vehicle back on course, and the driver made it across the river just as the surge approached. "You could see the water coming. That's the first time I'd ever seen it—I'd always heard about a wall of water. It was just like a wall of water just going down the river." The debris-filled wall was three to four feet high and rushed over the ford raising the water level much higher.

Some floods were so destructive that they washed out low-water fords and concrete crossings altogether. In July 1932 a devastating torrent destroyed the first river crossing. Hip Mengden watched some of the deadly force of the raging water from the safe vantage point of Slumber Falls Camp. Slumber Falls, another resort downstream from Waco Springs Camp, was operated by Mrs. Frances Lillie and her brother. It was a favorite spot for youth groups and had rustic cabins for accommodations.[17] For Mengden the rock bluffs above the river provided a ringside seat for the watery spectacle overflowing the banks below. "I remember in '32, the year of the flood, sitting on [the roof of] a cabin and watching a wall of water coming down the Guadalupe."

For the next year, with no real crossing place, automobiles had to drive the treacherous ford, but by June 1933 the county had built a new bridge. Writer Gertrude Rawson heralded the completion of the new structure:

> Since the disastrous flood of last summer, upper Comal County has had few visitors: the old bridge at this strategic point was destroyed and the only way of crossing the river was by the ford which was used before the automobile era. At a cost of $3,850 the county has now replaced the old concrete bridge with a new one, designed by Rudolph Jahn of New

> Braunfels, county engineer, and built by A. C. Moeller, contractor. A three-foot dam has been incorporated in the structure, supplying power for a private light plant and forming the lake used for boating and swimming at Huaco Springs Park.[18]

Comal County commissioners recognized the need for better and safer crossings along the Guadalupe River Drive. During the 1930s they awarded bids to have bridges built at all the crossings. Beginning in the spring of 1933 contractor Albert Meckel started work at the second river crossing with his bid to build a bridge at a cost of $1,639. In September 1935 he received the contract to construct a low-water bridge over the fourth crossing for $3,947, and two years later commissioners selected his bid of $3,710 for the building of the third low-water bridge.[19]

As a little girl Melba Meckel took an interest in her father's work and often watched construction crews in action. Building the low-water bridges required total commitment. Workmen did not go home each evening. They stayed on the job all week. "The first thing we would do," she recalled, "would be to set up camp, and we had huge tents set up for the carpenters to live in."

During summer vacation the Meckel family even lived in a tent at the job site. They cooked for the workmen. It was a real community effort. Everyone pitched in with whatever contribution could be made—men even went out hunting game—for food for the next day.

Constructing a low-water bridge was a complicated, tedious affair requiring hours of steady work. Melba recalled watching her father and his crew.

> They built it of course in sections. They would sandbag, and they built the forms. [Father] had a little generating plant because once you started pouring concrete—this was not some delivered by ready mix. This was mixing your own, getting the right combination of cement and sand and stone—mixed it right there on the job. And once you started pouring

> this section of the bridge, you did not quit.... The workers were using wheelbarrows to [haul] it, pour it into the forms.... It wasn't something you could do for eight hours and wait to continue the next day. So he would crank up his little generating plant to have some light and this would go until the wee hours of the morning until this section was completed. But everyone knew that before they started ... but they also liked this sort of thing..... I always thought the people who worked for my father were a great bunch.[20]

The work that the builders accomplished helped to open the scenic drive to even more tourers who, perhaps with map in hand, explored the meanderings of the byway. In spite of the area's growing fame, it still remained a peaceful, secluded place. "There was hardly ever that we would notice anybody come down the river on a boat or anything like that," Milton Kaderli commented. "They just never did that in those days."

Indeed it was big news when someone made a long river journey. On August 16, 1935, one youngster, Carl Saur, published a story, "Rowboating On The Guadalupe," in the *New Braunfels Herald*, detailing his adventure of rowing upstream on the Guadalupe the previous summer. He and his cousin embarked close to New Braunfels in a sixteen-foot rowboat with a flat tin bottom. As soon as they started they hit swift water that almost defeated them. "All along people had ridiculed our going upstream," Carl wrote. "Behind us on shore stood one anticipating our difficulty, probably waiting for us to turn back—and we felt like turning back. Flurried, we decided to continue, if only for a day."

But by the next day they crossed into Kendall County, and when they set out on the morning of the third day they turned back downstream. On the next day they floated back past New Braunfels and several miles downstream to their destination, the little hamlet of Clear Spring. They had traveled for a total of four days, having paddled about eighty-five miles of river and crossed 203 rapids. Carl wrote, "Although planned to be an inkling of a vacation, the cruiz [*sic*] neighbored nearer work; yet, we had enjoyable moments (shooting rapids, for instance) and do not regret having made it."

Recreation along the river mostly consisted of small private picnics. Longtime local newspaper editor Fred Oheim described the neighborly and

When the river was up cars had to ford the stream carefully. (Photo by Joe Sanders.)

A devastating flood had washed away the old ford. In 1933 a concrete bridge spanned the first river crossing and opened the Guadalupe River Drive to tourists once more. (Photo by Joe Sanders.)

respectful way in which people had fun at that time. "There was never a quarrel with people picnicking on the river banks. Most of the families had their little picnic lawns and if they were using them, you just didn't intrude." Otherwise, people asked permission to use the spot for the afternoon. "And of course, it was acceptable," Oheim added.

People cleaned up after themselves and were generally trustworthy. There was never any worry of vandalism or underhandedness. People felt safe and were free to enjoy the natural beauty.[21] Dorothy Kerbow remembered a fun-filled teenage excursion:

> One time when I was … a junior in high school, one of my friends … had an old Model T…. Well this would have been in about 1934…. By that time Model Ts were old and decrepit. But she had an old Model T that she had paid $15 for … and that old thing ran like crazy. It didn't have a top on it. It was a wide open two-seater. We played hooky from high school one time in that car, and we went out that river road. There were about four of us or five. It was full. And we went out there and we found a beautiful place. We just shucked our clothes off, and we went skinny-dipping all afternoon…. We felt safe.

The "fascinating and temperamental Guadalupe River" was one of the Hill Country's crown jewels, and the Guadalupe River Drive afforded motorists the opportunity to travel through some of the area's most beautiful scenery. Families returned time and again to enjoy the wild pristine countryside.

Chapter Nine

The Skyline Drives

> The winding hill-country roads, so isolated that deer or foxes or coyotes may easily be encountered, pass through the very typically Texan terraced hills, with cedar and cactus softening their crumbling limestone steps, and the white roadside dust made lovely by a successive profusion of wild flowers.[1]

The natural landscape of the country in northern and western Comal County blessed sightseers with breathtaking views along the lonely and sometimes treacherous hillside backroads. At the end of the Guadalupe River Drive, motorists could choose between several routes that skirted the hilltops.

Joe referred to the Hidden Valley road as "Skyline Drive." The byway offered a spectacular panorama of a farming area. In earlier days residents called it Bohemian Valley—named for the Czechs living in the region.[2] The zigzagging passage precariously hugged the side of a mountain. The motorist turned off at Sattler, went through a gate, and went along a road through pastureland and cedar. Suddenly the road narrowed dramatically with a sheer drop on one side and a wall on the other. It then dropped down into the valley. "But the road above is what I didn't like," said Laurie, "because it was so narrow." If a driver met somebody coming the other way, one would have to find a spot where he could back up and let the other car by. Joe became expert at negotiating the tricky route.

Laurie wasn't the only one who had a healthy respect for this route down the mountain. "It was kind of scary," Milton Kaderli remembered, "there were no railings or anything like that...."

The spectacular expanse of Hidden Valley. (Photo by Joe Sanders.)

Indeed the reward was well worth the trouble because scenery lovers heralded the Hidden Valley road as offering one of the most beautiful views in the county. After wayfarers struggled along a narrow rocky trail, all of a sudden their next turn revealed an expansive valley patchwork of golden fields.

From Sattler, in addition to the river road, an alternate route to New Braunfels was the Prairie Road. It followed roughly the same course as Farm Road 306 would in later years, and different stretches went by various names like the Hancock Road or, earlier, Marienthal Road, named after Marie Hesler, the wife of early county settler James Ferguson. Hancock was another scenic village in the hills, and in the spring nature greeted drivers with a full spectrum of color from budding trees and wildflowers. Green hills rose to the sky behind endless meadows of bluebonnets.[3]

Melba Meckel made many trips on the Prairie Road. In fact, her father taught her to drive on it, but her early memories of the route came from riding to town with her grandfather. His old green, temperamental truck had solid rubber tires and predictably dictated the pace on the rough and rolling road. "You had to go up Baetge's Hill," she recalled. "You could get just so far and the vehicle would get too hot and you had to stop." On her grandfather's signal, Melba hopped out and wedged rocks behind the tires to prevent the truck from rolling backwards. While the engine cooled, Melba and her grandfather admired the countryside. "He always knew when to start that truck up again," she commented. They went on their way down the bumpy course. During the 1930s her father Albert Meckel helped to rebuild the roadway, smoothing it out and extracting flint rock with the assistance of WPA labor.[4]

Another scenic but dangerous road was the Devil's Backbone Skyline Drive in northern Comal County. One feature writer described the Devil's Backbone as demonstrating "the creative genius of Nature."[5] The awesome formation appeared as a series of rugged ridges. Beyond they harbored the picturesque village of Wimberley in Hays County. Drivers atop the winding steep could spy cedar-covered ranges on both sides, but like all the hilly byways of the county, it too claimed its own overheated car engines. Many motorists even backed up part of the grade in an attempt to sustain more

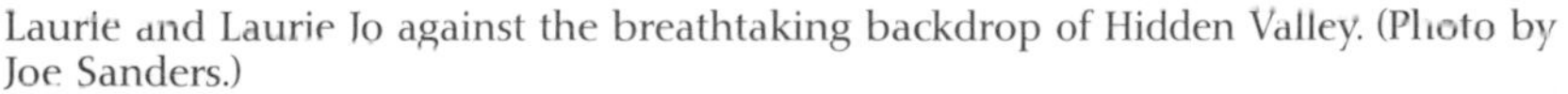

Laurie and Laurie Jo against the breathtaking backdrop of Hidden Valley. (Photo by Joe Sanders.)

The Baetge Hill area in Comal County. Stretches of this road have been known by different names: Prairie Road, Hancock Road, and Marienthal Road. (Photo by Joe Sanders.)

power, but the rocky path had no shoulders, making for an even more nerve-wracking situation.[6]

Whenever Dorothy Wimberley Kerbow and her family went to Wimberley via the Guadalupe River Drive, that route meant that they had to go over the Devil's Backbone. "It was terribly dangerous!" she exclaimed. "If a big rock had fallen down into the road, and if my father had to drive around that rock and drive near the edge, my mother would make me and my brother and sister get out and walk."

To say the least the drive was thrilling, and it required a fair amount of skill and finesse to maneuver. Farther northwest it wound its way to Fischer's Store.[7] The little community of Fischer's Store had been a Hill Country stopping point for years. In the 1850s Herman Fischer, Sr., established a mercantile that originally stocked a large assortment of ranching goods including farm machinery, hardware, and groceries. His enterprise also included a saloon and post office (which designated the spot Fischer Store in 1875, dropping the possessive from the name). The U.S. Postal Service criticized the presence of a saloon as a clear violation of postal regulations, but since Fischer was the only person available to serve as postmaster of the little village, post office inspectors agreed to overlook the transgression in exchange for a "drink on the house." By the age of the motoring tourist, the business consisted of a post office, general store, and filling station.[8]

The hamlet, which had been a favorite resting place for early travelers and horse-drawn freight wagons, was still a welcome respite for drivers. On one Sunday afternoon, while Joe logged the store's position and the nearby mileage for his map, the rest of the family enjoyed cool drinks and snacks from the mercantile. Cattle lazily roamed in the nearby pastures and goats ranged free along the roadside, grazing in the tall grass, while young billies frolicked around the storefront. Not only was the store popular, the community also had nine-pin bowling and Saturday night dances for the enjoyment of farmers from several surrounding counties.[9]

The hills outside of Wimberley in Hays County. The hill, part of the Leath Ranch, was called Eagle Cave Mountain. (Photo by Joe Sanders.)

Several roads emanated from Fischer's Store. Beyond the village one of the routes eventually led northwest to Blanco, but the journey took time as the byway seemed to wind around every cedar tree. In pioneer days travel could be very onerous. The *San Antonio Express* reported:

> In rainy weather the roads were awful, even bad in dry weather. In dry weather the ... freight [wagon] failed to meet expectations and missed schedule by hours, sometimes by days. It finally got hopelessly stuck in a mudhole.[10]

Some improvement came in the early twentieth century. In 1924, road crews worked on the section called Pfeuffer Hill on Fischer Store Road. The men cut down about four feet from the top of the hill and leveled the roadway gradually over the length of 180 feet until the way was flat. Even though they smoothed out the course it was still only eighteen feet wide. The 1930s saw more progress, and the *Express* proudly reported, "Autos now have all-weather roads radiating in four directions from Fischer Store. No mudholes, but plenty of bumps."[11]

Motorists who ventured on toward Blanco from Fischer's Store came upon the charming pastoral valley of the Little Blanco River, and from there they could cross into Hays County on to the private park of the Narrows on the Blanco River.

Heading south from Fischer's Store tourists crossed the Guadalupe River and reached Cranes Mill Road. Cranes Mill, another country burg, gave drivers a chance to refuel and rest before proceeding through the rolling hills.[12] Continuing south, adventurers could visit Startz Hill, which, at an altitude of over 1,400 feet, was the highest point of Comal County. Eventually property owner Carl D. Allen would donate land on the hill for a roadside stop called Allen Park.[13] Startz Hill granted vast views of the countryside. To the west, sightseers could see the village of Smithson's Valley. The summit provided a clear survey to the northwest over the ranches along Rebecca Creek and straight through to the profile of the Twin

Sisters—two mountains sitting side by side above the horizon in Blanco County. Startz Hill displayed its ancient seabed origins and provided a "glaring example of nature in erecting a rampart to prevent the receding

The rocky cliffs of the Narrows on the Blanco River. (Photo by Joe Sanders.)

waters of the sea in interfering with her creative scheme in the valley below."[14] The natural uplift provided explorers with interesting fossils and water-formed rocks. Along the hills the Sanders family found numerous geodes. Some were perfectly round. Others had been split open and revealed their crystalline centers.

Some of little Laurie Jo's fond childhood memories were of the explorations the family made during the days when the landscape was a more accessible place. In those years, "they didn't have all the 'No Trespassing' signs up. You could go beyond a fence and climb a mountain if you wanted to." Along the isolated roadsides, landowners didn't mind a little innocent exploration. People trusted each other. No one ever chased the family away. "There wasn't enough [exploring] going on that they had any cause to worry," Laurie Jo explained.

From Cranes Mill Road motorists could turn onto Bear Creek Road, which, according to the scenic map, traversed "possibly the wildest and most rugged section of the county." Text on the back of the map added, "There's a good chance of seeing wild animals through here." Wildlife abounded in the pristine area, including armadillos, rabbits, deer, wild turkeys, foxes, coyotes, and bobcats. Ranchers even glimpsed an occasional cougar in the remotest areas. Towards dusk it was common to spy opossums, raccoons, skunks, or ringtails.[15]

Bear Creek was a favorite destination for the Sanders family, and so they traveled the road often. Here too, the land was rich with arrowheads and other artifacts of a bygone era. In the vicinity of Bear Creek the family discovered a large mound that they suspected may have once been an Indian burial ground.

Comal County had a hidden archeological wealth of relics—a fact proven in the summer of 1936 when a road crew unearthed a skeleton estimated to be 4,000 years old and representative of a race of mound builders.[16] They made the find on the outskirts of New Braunfels near Klappenbach Hill on Fredericksburg Road. Melba Meckel Roth recalled, "When it was badly in need of repair, father was contracted to fix the road,

Mountain scenery in Comal County. (Postcard photo by Joe Sanders.)

and [when they started to grade], they discovered some Indian artifacts...." Many of the objects went to citizen Albert Nowotny who ran a popular private museum on San Antonio Street called The House That Jack Built.[17]

The *San Antonio Express* commented:

> Comal County is rich, both in prehistoric and ancient history. Each year sees additional finds made of fossil bones of giant reptiles, sea fossils, and the relics of primitive man. Nature has been also more than generous in placing her bounties and it is now believed by geologists that a vast subterranean lake lies hundreds of feet beneath the surface of the area between San Antonio and Austin.[18]

Exploration rewarded many rock hounds, and some of the treasures were in a small museum in the courthouse, to which the road map directed tourists, stating, "If the geology of the region interests you, let us suggest that you inspect the fine mineral and rock collection of A. M. Fiedler, County Superintendent, at the Courthouse." Many relics were eventually displayed at the town's official museum, the Sophienburg Memorial Museum, which opened in 1933.

The Sunday motorist had a fine choice of beautiful vistas. "Harsh as these hills may appear in the glare of the foreground, they blend the distance into an indescribably rich panorama," Gertrude Rawson wrote, "with a deep wool tapestry effect in strong blues and greens."[19]

• • • • Chapter Ten • • • •

Nature at Work: Secret Places, Caves, and Creatures

Hill Country sightseers driving along high winding ridges saw only the wonders above ground. But the topography, with its rolling hills of limestone, also harbored a hidden land of quiet grottos and carved-out rock formations. The "subterranean vaults"[1] in the terrain, eons in the making, provided an underground landscape, different yet as breathtaking as above.

Tourist businesses, many of them family-run, offered close-up exploration. "If you want to see a cave in its natural state, not prettied up for commercial purposes," the Centennial road map suggested, "Ed. Heidrich, the Snake King of Comal County, will gladly take you through Brehmer's Cave on Saturday or Sunday." The small operation outside of town was a popular local stop and drew mention in the *San Antonio Express.*

> ...nature evidently overdid herself in this part of Texas in providing forms which today are viewed with awe by the tourists who visit here. In the interior of Brehmer's Cave south of New Braunfels we view the stalagmite formations which show where the land masses failed to knit properly and to form hollow channels in the subterranean land masses. Many caves not even marked are to be found within a radius of 10 miles of the Brehmer Cave. All of these caves provided underground passages

of Comal County and a test made of voice vibration from the Brehmer Cave was heard plainly in a cave two and one-half miles south of New Braunfels.[2]

The Hill Country was honeycombed with cavernous chambers, the result of water percolating through the limestone over millions of years. Joe, ever the amateur caver, loved to explore for subterranean wonders. When he heard of a cave off the Guadalupe River Drive, he had to investigate it.[3] Then, full of enthusiasm, he went home to tell Laurie and take her to the spot. "I had to go look at it. Whether it was a natural cave or whether it had been perhaps opened up by the Indians," she described, "it was more like a shelter, like they might have taken shelter there from a storm." The site, no doubt repeatedly carved out by countless river floods, contained arrowheads and other Indian artifacts.

As the map described, the Texas Hill Country offered "weeks of exploration in secret places where dark waters reflect the feathery moss-draped forms of giant cypress trees." Scenic marvels also extended outside Comal County into the bordering counties, and Joe's road signs directed motorists to nearby attractions.

One interesting destination near the hillside village of Wimberley in Hays County was Jacob's Well—a deep and mysterious underwater cavern. The feature generated so much curiosity that in the 1930s brothers Earl and Roy Swift of San Marcos fashioned homemade diving gear to explore the "well." They dispelled the old legend that the watery chamber was the flooded shaft of an abandoned silver mine.[4]

The well spring, actually the head of Cypress Creek, provided "a steady column of cold blue water welling up through a vase of solid rock pearly grey and fluted as regularly as a jelly mould."[5] Jacob's Well was a favorite swimming hole, and young Dorothy Wimberley Kerbow enjoyed the crystal artesian waters. "It was like a fountain. The whole thing just surged. You could dive right into the center of it, and it would just bring you right back up. Oh, that was a glorious feeling—what a sensation!"

Young Joe peers out of a cave in Panther Canyon near Landa Park. The subterranean features common to the Texas Hill Country were a constant fascination to him. The picture was taken about 1917, shortly after Joe arrived in Texas. (Laurie P. Sanders Collection.)

Watery splendors also awaited the weary traveler near the hamlet of Kendalia west of Comal County in Kendall County. Edge Falls on Curry Creek provided a peaceful cool hideaway, and the Sanders family visited the privately owned park. Joe and Laurie had made their first trip in 1929. They climbed though a barbed-wire fence to get there, then stopped at the owner's house to ask permission before venturing to the falls.[6]

Rickety iron stairs led down to the creek and into an enchanting lush green glen. Springs gushed out of the rocky cliffs, and stepping stones were nestled amongst ferns, watercress, and velvety mosses. The waters of Curry Creek cascaded from an overhanging cliff down into an emerald pool. Laurie followed the stone path along the creek's edge and wended her way to the waterfall. A grotto, naturally carved into the cliff behind the waterfall, provided a haven in which to stand and watch the curtain of cool water.

A few years later when Joe was compiling the map, the family returned, and little Laurie was equally charmed by the place. "It was absolutely beautiful. We sat up above where the falls came down and dabbled our feet in the water."

The numerous springs and limestone creek beds throughout the surrounding region hinted at the existence of a network of underground scenic wonders. Landowners sometimes tried to take advantage of their locales by establishing cottage tourist enterprises. The Sanders family visited one such small operation—a cavern in eastern Kendall County. Laurie remembered:

> [Joe] had heard about one called Fairy Caverns.... So we went to the place where he had been told there was this cave, and he met the man who owned the property.... He hoped someday maybe to open it up and really profit by it.... He had built a stairway down there carved out of earth, and he had put a little railing on one side so you could go down without breaking your neck. It didn't look very safe to me, but I went down there anyhow.... He didn't have it lighted at that time.... We had flashlights, and ... the sun was showing a little light down into the opening you know.... Joe put it on the map hoping to generate interest....

Laurie Jo was very impressed with the formations of caramel color shading into burnt orange. "Very, very beautiful," she marveled. They walked along a pathway that led to a small pool as green as an emerald. "I think there was a little ledge or small steps beside it," she said, "and you could look down into that beautiful green pool, and you could hear the water dripping too."

In 1932 businessmen in the area attempted to launch a large-scale tourist venture called Cavern City, and Fairy Caverns was to be a part of it. The *San Antonio Express* reported the development of five big caves on 1,000 acres. Real estate developer E. N. Requa purchased the property and estimated spending more than $100,000 to develop Cavern City Resort. His plans included constructing a skyline drive, excavating and lighting five caves, and building tourist cabins. One of the attractions would be named the Cavern of Palaces. Another tour would include boat rides on an underground river.[7]

The proprietor at Fairy Caverns did eventually print brochures advertising it as a "miniature Carlsbad" and "fairy palace." By that time the cave had electric lights, and the owner charged an admission price of 25 cents for adults and 15 cents for children. "The Cavern is one of many such in Cavern City," the ad read, "a tract which has been valued at $200,000 for its scenic beauty and other attractions...." Formations had colorful names like the "Frozen Cascade," "Cactus Forest," and "Witch's Firepot."[8]

The grandiose plans for Cavern City never materialized, however, and the cottage industries remained small and eventually closed—victims of the Depression. Cascade Caverns, another attraction in Kendall County, did continue operation and was also included on Sanders' 1936 map.[9]

One unique cave provided entrepreneurial operations dependent not on tourism, but on its valuable natural resources. Bat Cave north of the hamlet of Bracken in southern Comal County was the site of a fertilizer enterprise. Its huge population of millions of bats resulted in the production of massive quantities of bat guano.[10]

"They were gathering guano from it and selling it," Laurie recalled. "They had made ... an elevator shaft down from the top, down deep into the cave, and they'd let these little cars down in there, and they'd fill it with guano and then reel it up." Joe bought a large burlap sack of guano for the plants and shrubs around the Sanders house. The guano was a potent fertilizer due to its high nitrogen content, and the substance had figured prominently in the previous century during the Civil War when workers mined and processed guano to make saltpeter for the production of gunpowder.[11]

In addition to the mining enterprise, the cave attracted the curious to witness the evening spectacle of myriads of bats making their nightly exodus into the sky. Awesome dark clouds of bats flew out. Their night flight was highly beneficial, because the creatures consumed massive amounts of insects.

The entrance to Bat Cave appeared "like a dugout bowl"—a big black depression in the earth. Joe and Laurie Sanders stood on level ground on the rim above the opening and stared down into that dark abyss. Laurie wouldn't go in for anything. Joe, on the other hand, with his undeniable sense of curiosity and adventure, felt differently. "Of course he had to go down there with a flashlight," Laurie sighed, "not the time that I was with him, but some other time when he was there."

When Joe did venture into Bat Cave, his light faded into the endless blackness. It was too dark to discern anything. He did come across various subterranean creatures, albino lizards, cave crickets, and toads. Fortunately, the bats nesting deep in the interior were not to be seen, and, failing to penetrate the relentless darkness, he ended his exploration.

Joe was always watchful of any telltale signs of underground openings while on his road treks. Once, while driving down Purgatory Road in the northeast part of Comal County, he noticed a clump of trees in the middle of a field. Suspicious that there was some sort of hidden cave or rockery nestled there, he parked the car, and, while the family and Laurie's visiting mother waited, he took off to investigate the spot. He confirmed his suspicion when, peering through the trees, he saw a large rocky shelf. Beyond was an opening, about four feet across, in the shape of a half circle.

With flashlight in hand Joe crawled into the cave to look around. He explored a little and got the lay of the place, but as he turned to crawl out, he came face to face with a rattlesnake coming in. Fortunately Joe sometimes carried a small pistol on his adventures, and he was thankful to have it along on this occasion. Unaware of the danger at hand, Laurie and the family sat in the car.

> All of a sudden we heard a shot, and at the same time a rattle—rattlesnake—so loud that I believe you could have heard it a mile away! I knew it was a rattlesnake, and we didn't see him [Joe]. We knew that

> he'd gone down into the bushes there, so I got out of the car and went through the fence and went running over there, just in time to see him looking out of the opening of a cave.... Right outside of it was a large flat rock just like a shelf coming out from it, and on there lay this great big snake—and he was a huge one. He looked like he'd be about five or six feet long ... and he was rattling, sort of sidewinding there, wanting to get back in the cave. And of course Joe WAS in the cave looking out.... He'd shot it once with his pistol, and shot it again, before I got there. He hit the snake but didn't kill it. And as it slid down one side of the cave, the same time Joe came out! So that was a narrow escape.[12]

Sometime later while driving down Purgatory Road he stopped and looked at the cave but never went in there again.

Snakes, of course, were as prevalent in the countryside as other wild critters, and the road map included small sketches of coiled up serpents along with the other animals. In later years Chamber of Commerce members asked Joe to remove the snakes from the map, because they felt the creatures conveyed the wrong image for Comal County.

Nevertheless the reptiles were a natural part of the landscape, and Joe was as fascinated with them as anything else. "In the summer evenings he would drive out," Laurie recalled, "and it almost seems like nearly every evening we would see a rattlesnake crossing the road." He liked to catch the snakes and had the necessary equipment—a snake stick and a wire mesh cage. The snake stick was a long sturdy staff with an adjustable leather strap attached that could loop around a snake.

One particular hunt, though dangerous, had its humorous moments. Joe cornered a rattlesnake in a shallow ditch by the side of a country road. Holding the creature down behind its head with another large stick, he attempted to snare it with his snake loop and enlisted Laurie's help.

He asked her to come and hold the snake down with the stick. "There I was holding him, pressing down on that stick 'cause he was a big snake and tried to get out from under it and couldn't.... It was a crazy thing to do.

I was putting all my weight on that stick." Joe had already caught another snake—a whip snake—and he didn't want to put the rattler in the box with that snake. So he handed the whip snake to Laurie. "I knew he was harmless, but just the idea of it," Laurie exclaimed. At that point a car came by. The passengers must have thought the Sanders were two of the craziest people they ever saw.

Joe's peculiar hobby had some benefits though. He sold the snakes to local nurseryman Otto Locke. The Locke family had operated a gardening business in New Braunfels since 1856, when Otto's grandfather Johann Joseph Locke first started the nursery. Otto also kept a "fine collection of wild animals, snakes and birds" according to the 1936 road map. His interesting zoo included ringtails, ocelots, raccoons, monkeys, peacocks, a coatimundi, a boa constrictor, and even a pet puma that would ride in the back seat of Locke's car. The nurseryman paid Joe according to a snake's weight. Of course, as Laurie observed, "A snake is like a fish. They don't weigh as much as you might think." But Locke paid 50 cents per pound for a snake, and rattlesnakes were usually the heaviest. A good-sized rattler weighed about six pounds and therefore brought $3. A buyer in San Antonio then purchased rattlesnakes, which were milked for their venom in order to produce antivenin serum for treating rattlesnake bites.[13]

Joe also went on snake outings while chauffeuring Mr. Dittlinger. Hip Mengden remembered, "He'd see one and boy we'd come to a stop. He'd peel out with a stick with a loop on the end and he'd catch them." Other Dittlinger grandchildren, Maria, Carl, and Loretta Liebscher, witnessed the hunts as well and remembered their sense of amazement. Maria Liebscher recalled the times when Joe and her grandfather stopped for a serpent adventure. "And of course we kids went right along with it. We thought it was something wonderful. He always had his snake loop along."[14]

Though Joe looked for rattlesnakes, occasionally he spied other snakes—bull snakes that let out a big roar or hog-nosed snakes that puffed up menacingly and then feigned death. A chicken snake somehow maneuvered himself out of the snake box once and got loose in the car. Joe found it curled up in the springs under the back seat, and after removing the whole seat, he finally caught the snake and "oozled" him out.

Laurie had some snake experiences of her own. When she was a teenager she hunted copperheads with her rifle along the Guadalupe River in Victoria and became a crack shot. Having never really feared the reptiles, she recognized the benefits of many of them as pest controllers, but that familiarity almost got the best of her. In July 1937 while on a visit in Victoria, she caught what she thought was a little harmless indigo snake. The reptile was actually a freak of nature—a black coral snake. While carrying it home, she eased her grip ever so slightly and the snake latched onto her finger. Waves of pain began almost immediately, and she was bedridden for almost two weeks. Luckily Laurie recovered, and, having topped even the largest of Joe's rattler catches, for a time she became somewhat of a celebrity to snake experts. Several area newspapers covered the story, and the case attracted the attention of noted herpetologists Howard Gloyd,

The rare black (melanistic) coral snake discovered by Laurie Sanders was believed to be the first ever found in Texas up to that time. This photo taken by the noted herpetologist Howard Gloyd of the Chicago Academy of Sciences appeared with his article about the biting in the scientific journal *Herpetologica*. (Photo by Howard Gloyd.)

Colonel Martin Crimmins, Raymond L. Ditmars, and Dudley Jackson, as well as personnel at San Antonio's Witte Museum. She also received letters from all over the country from buyers who mistakenly took her to be a snake handler in the snake trade.[15]

Snake stories aside, for the most part the family's favorite adventures consisted of scenic excursions along the wild winding roads. The backroads enabled all kinds of exploration, and Joe was always eager to traipse through new country. They were out on a Sunday drive in the southern section of the county one mild January afternoon. Joe couldn't resist the lure of a particular meandering backroad. Following the byway until it narrowed, he finally parked the car when the path became practically impassable because of all the tangled brush. Joe was just naturally curious about the surrounding countryside, and he said he "liked the looks of it." So, armed with a Kodak camera, he trekked off into the thicket. Laurie, suffering from a toothache, stayed in the car. "He was gone for quite a while, and I thought, 'Where in the deuce is he?' Finally he came back, and he said,

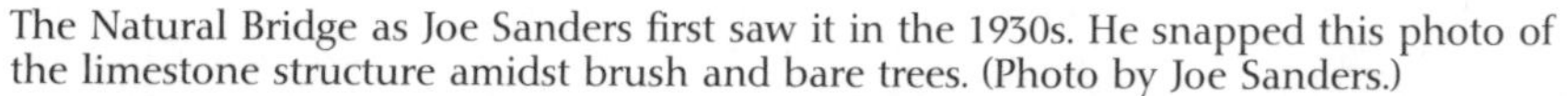

The Natural Bridge as Joe Sanders first saw it in the 1930s. He snapped this photo of the limestone structure amidst brush and bare trees. (Photo by Joe Sanders.)

Collapsed rock sides and fallen boulders about the hourglass-shaped bridge provide the telltale signs of a sinkhole and hint of subterranean chambers. Scads of varied arrowheads and stone tools were visible on top of the bridge. (Laurie P. Sanders Collection.)

'You know there's a natural bridge back here!' I felt like saying 'So what,' and he said, 'Come and look at it.'"

Joe finally persuaded Laurie to tromp through the brush. Sweeping away the naked branches of bushes and brambles, she caught a gleam of winter sunlight angling through the bare trees, casting spindly shadows across this natural bridge. The limestone structure, about sixty feet long and twenty-five feet high, spanned a gully, and below one end tumbled up rocks and earth covered what Joe surmised to be the buried entrance to a cave. Arrowheads were abundant, and they imagined that the spot was once an important Indian refuge.

The Wuest family owned the property, and Joe asked their permission to put the bridge on his scenic road map. They agreed, but in later years they asked him to take the sketch off any future maps, because every Sunday afternoon they were hounded by people wishing to view the bridge.[16] Obviously the road maps caught the attention of numerous sightseers.

So it was that every excursion meant the possibility of exploring secret and beautiful places during a time when the country was freer and people were more trusting just as long as the curious left the scenic wonders undisturbed. Families picnicked and stopped to snap pictures destined to immortalize an afternoon in the family Kodak album. While no one had much money to spare, a cool haven or majestic vista was always within an hour or two. With a few gallons of gas and a sack lunch, drivers followed the arrowed signs at the nearest crossroads—off to some scenic byway.

. . . . Part 3 . . . .

AT A CROSSROADS: THE CHANGING FACE OF TRAVEL

"...motor on

for a while..."

· · · · Chapter Eleven · · · ·

Scenic Souvenirs

When the Sanders family went to the Texas Centennial Exposition in Dallas in the fall of 1936, the festivities were winding down. The last days of the exhibits and shows capped off a celebration that reflected a climate of new ideas and bold innovations inspiring wonder and imagination in all visitors. The Centennial brought about major increases in tourism, and as part of the state's effort to show its best face, the highway department's work for better byways and roadside parks made driving easier. Statewide improvements in turn encouraged individual communities to expand their publicity to cater to a growing travel industry that would last beyond the Centennial.[1]

In Comal County, the success of Joe's road-marking program helped to establish the reputation of the area as a scenic destination and to encourage a prosperous new tourism enterprise that aided the local economy. In October 1936 in an attempt to snag even more drivers and advertise the town of New Braunfels along the main thoroughfare, the Chamber of Commerce proposed to erect its own markers—a "series of twenty artistic sign boards on Highway No. 2 between San Antonio and Hillsboro." City officials did not fully carry out their initial plan, but in February 1937 the board of development did place ten signs along the highway from New Braunfels to Waco at an expense of about $300.[2]

Comal County's friendly roads earned deserved praise from motorists across the state and the nation. Joe and the Legion received thanks from satisfied sightseers. One letter dated October 9, 1937, from Alexander L. H. Darrogh of Chicago, was printed in the *New Braunfels Herald* on October 15, and represented the sentiment of many travelers:

> Mrs. Darrogh and I wish to express our thanks to your Post for the many road signs (accurate, too), for the road map printed by your Post, which a local garage gave us when we made several trips up from San Antonio to the country near your city.
>
> In particular we enjoyed the Devils Backbone, Hidden Valley, Wimberl[e]y, the Guadalupe River canyon....

The Hill Country left a lasting impression on wayfarers who ventured over the winding backroads, and Joe looked for additional ways to promote Comal County's beautiful landscape. He envisioned a series of postcards that would serve as scenic souvenirs for tourists and townspeople alike, and he sought the assistance of local photographer Otto Seidel.[3] His son Rudy Seidel recalled, "Mr. Sanders had the idea to take the pictures and make postcards out of them to ... promote the area.... He came to my parents with it, and they thought it was a good idea so they just kind of worked together."

The Seidels had established their studio in New Braunfels in the 1920s. During the 1930s they operated their business on West San Antonio Street in the second story of the Hinman building, above Vollmar's five-and-dime store and across from Henne's Hardware. Seidel photography became an intimate chronicler of the town. Their works captured the significant events in the lives of citizens through family portraits, wedding photos, and funeral pictures. Seidel also engaged in field jobs, such as photographing automobile accidents and train derailments for the benefit of interested groups such as the police, the railroad, the local newspapers, and insurance companies.[4]

When Joe Sanders came to Seidel with his postcard idea, Seidel wanted to contribute to the project, but with a busy photographer's schedule he could hardly travel the countryside. "My father didn't have the time to go up in the Hill Country on Bear Creek and River Road and all those places," Rudy observed.

Therefore, while Seidel took pictures of Landa Park and other points of interest around New Braunfels, Joe photographed scenic areas throughout

comfort food

Stew Master!

One big cast-iron pot and a heap of down-home flavor

BY TIM BENDALL, PRINCE GEORGE, VIRGINIA

PHOTOS BY CHARLES "STR

YOU MIGHT SAY I'VE GOT stew in my blood. It started with an article in my local paper a few years back about Brunswick Stew—one of those great old Southern dishes born of thrifty ingenuity. The story goes that some hunters were out on a long trip, and when suppertime rolled around, the fellow doing the cooking didn't have much to work with—some stale bread, potatoes and onions, a few squirrels they'd caught that day. This intrepid hunter-cook seasoned the stew with lots of salt and pepper, and set it to simmer over the fire. It's a time-honored culinary institution now down here in southeastern Virginia, where I was born and bred. These days, we make it with chicken instead of squirrel, and a lot more vegetables. Like all folk traditions, it has evolved. It's a complete meal in one dish. Around here, you might see huge cast-iron pots holding 95

FAMILY STYLE *Tim with his wife, Marilyn, and son, Gary* gallons of the delicious stew, bubbling away at family reunions and church suppers. It's almost as big as barbecuing. And, for my money, better.

I've known of Brunswick Stew my whole life. My dad used to make it on special occasions. Always on July 4, when there were so many tasty vegetables fresh from the garden—corn and tomatoes and lima beans—all cooked over an open fire. I regretted that I never took down the recipe before he passed away. Sometimes he'd let my brother and me help build the fire or take turns stirring the stew with the big maple paddle. Something about making that stew brought us closer. It's the kind of cooking men can bond over.

So when I saw that article, it brought back lots of good memories. I noticed something else it said: A man named Chiles Cridlin was a local master of the art of Brunswick Stew. Chiles Cridlin? Surely there could only be one. It had to be the same man I worked with at the furniture manufacturing company where I'm a quality-control engineer.

It all began with an article in the local *paper a few years back about Brunswick Stew.*

COME AND GET IT! *Tim made Brunswick Stew for a family gathering this past fall.*

"I read that article about Brunswick Stew," I told Chiles later. "I didn't know you were a stew master."

He laughed. "I can't take the credit. That's my son. He's the stew master!"

I told Chiles about my father, how he used to make Brunswick Stew, how I had great memories of cooking with him. "Just wish I'd gotten the recipe," I said. He put me in touch with his son right away, who was happy to have a new recruit to the Brunswick Stew cause. I discovered from him that the Virginia Foundation for the Humanities has an apprenticeship program for an aspiring stew master. I signed right up.

It's still funny to me how that one newspaper story awakened my passion for this stew. I was on fire. I wanted to learn everything there was to know. I was determined to become a certified stew master—just like Chiles and four of his buddies in the "Proclamation Stew Crew": John Drew, Lonnie, Phil and Rodney. Soon, the six of us were cooking Brunswick Stew around the state and even around the country. We love to make stew for charity events. It's a great feeling

GOT A STORY ABOUT A RECIPE THAT'S CHANGED YOUR LIFE? *Write to Guideposts, "Comfort Food," 16 East 34 Street, New York, NY 10016, or submissions@guideposts.org.*

to do something you love that helps out good causes. Sometimes I think our wives and kids may worry we're too focused on this stew business. But they understand what it means to us.

There's some argument about where the first Brunswick Stew was made. Of course, I believe it was Brunswick County, Virginia—just two counties away from me. Some folks down in Brunswick, Georgia, are inclined to disagree! But that's beside the point. It's more than just a dish; it's something that connects families and friends through generations. I think back to those hunters who first came up with the stew; they could've been my forefathers. Like I say, it's like it's in the blood. ■

For more on this story, see Family Room.

Brunswick Stew

THIS STEW MAINTAINS ITS LEGENDARY STATUS AS A SYMBOL OF HEARTH AND home. The thick consistency is great for those cold fall or winter's nights.

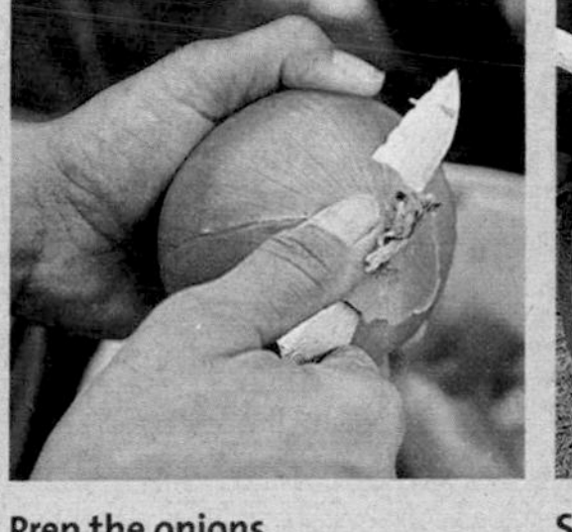

Prep the onions.

Stir in crushed tomatoes.

Time for corn. A meal all in one!

- **5 ½ pounds de-boned chicken (thighs are best)**
- **6 ounces white meat (fatback), ground or chopped**
- **4 pounds white potatoes, cut up French-fry size**
- **2 ½ pounds yellow onions, chopped**
- **¼ ounce black pepper (to taste)**
- **¼ ounce red pepper (to taste)**
- **1 ½ ounces salt (to taste)**
- **1 ½ ounces sugar (to taste)**
- **1 ½ quarts crushed tomatoes**
- **2 ½ quarts small green butterbeans (lima beans), drained**
- **1 ½ quarts white shoepeg corn, drained**
- **1 stick margarine**

Prepare potatoes and onions ahead of time so you can stir continuously (constant stirring is necessary for the thick consistency). Put chicken and white meat in pot. Cover with water; bring to boil and cook until chicken starts to come apart. Add potatoes, onions and 1/4 of seasonings. Bring to boil again and cook until potatoes are soft. Add tomatoes and 1/4 of seasonings. Bring back to boil and cook 5 minutes. Add butterbeans and 1/4 of seasonings. Bring to boil and cook until butterbeans are soft. Add corn, margarine and rest of seasonings. Cook 10 to 25 minutes, then serve (Feeds 20).

the archivist

BY VAN VARNER

How a secret giver helped start Boys Town

FATHER FLANAGAN'S ANGEL

I'VE ALWAYS THOUGHT THAT FATHER Flanagan's Boys Town was the essence of what we like in a GUIDEPOSTS story. It's a place where thousands of troubled boys and girls have had their lives turned around. For example, in the December 1993 issue of the magazine, we ran a story by Ron Dennis, who came to Boys Town as an adolescent after doing several stints in jail. There he learned to "make his life count for something" and went on to the Marines before returning to Boys Town as a teacher. But behind his inspiring story there is an inspiring story about how this now-famous institution got its start.

Back in 1917 Father Flanagan envisioned a place where troubled kids could learn self-governance and get skills they could use to earn a living. From his work with homeless children on the streets of Omaha, Nebraska, he saw that sending boys to reformatories under lock and key, where they associated with hardened criminals, only taught them to be criminals. He proposed to break that cycle by making the boys responsible for themselves and each other. He would start a new kind of orphanage in a house he had found in Omaha. All he needed was ninety dollars for the rent. He was a Catholic priest and had no means himself. What he had to find was a donor sympathetic to his revolutionary vision.

That's when he turned to Henry Monsky, a prominent Jewish lawyer in town. Although the two men were of different faiths, they had a passion for the same causes. Over the years they both worked hard for the prevention and control of juvenile delinquency. Yes, Monsky would agree to putting up the ninety dollars, a sum worth considerably more in those

The New Braunfels plaza downtown ca. late 1930s. The courthouse is on the left. Note the signs with highway numbers and mileages. The plaza had two-way traffic through the late 1940s and into the 1950s. (Otto Seidel Collection.)

Comal County. An amateur photographer, he had always been interested in the intricacies of cameras and the art of picture taking. Seidel gave him a large fold-out handheld camera. "It was called a postcard camera," Rudy said, "and it made a postcard sized negative...."[5]

So Joe traveled the county and captured majestic vistas along the Guadalupe River Valley, the rugged cliffs of Bear Creek, quiet reflections on emerald waters, and meandering country roads. He took many of the photos during the late 1930s and also incorporated some pictures he had taken in previous years. He and Seidel shot approximately fifty views of the county and city.[6] Joe then performed the necessary legwork to generate community interest. "He went out to all the drugstores and the five-and-dime stores ... the hotels ... anybody that would sell postcards," Rudy explained. "He went out and took the original orders for them."

Then Seidel printed the postcards at or below production cost. The massive job was relegated to nights, and the whole family engaged in the undertaking. Young Rudy was no exception. He and his older brother Edmund worked past 1:00 A.M. for many sessions.

Quiet reflections on Bear Creek. (Photo by Joe Sanders.)

> After six o'clock, after the studio closed and my parents would eat, then we'd go down with them, and we started printing … postcards…. One night my brother would help, and the next night I would help. You see we were still going to school. But my parents did this every night for weeks till we had all of them made…. My father would print the post-cards. My mother would develop…. We put bigger bulbs in the printer to where the exposure time would be shorter, and he would [expose the negative] just a split second and then he would throw it in the developer, and my mother had both hands in there, because he'd throw them in there fast and furious….

They produced the cards by contact printing directly off the negative onto a stiff postcard paper. In the development process, once Mrs. Seidel ascertained that a print had reached its correct density, she put it in the short stop to end the development of the image, and then one of the boys put the prints into two sets of fixing solutions and eventually into the wash water. "It was a job. It was mass production," Rudy commented. "We'd put out a lot of postcards in one night."

The exhausting job dragged on as the family worked a succession of late nights; some nights were more memorable than others. On one occasion after Rudy and his mother had finished around 1:00 or 2:00 A.M., they waited in their car and relaxed while Mr. Seidel closed the studio. Fatigued, they lost track of time, but when they realized that he had never appeared, they became concerned and went back to find the studio locked and Seidel gone. In a panic, the family ran to the police station nearby and reported a kidnapping. The police advised them to go home and wait for any news, but upon arrival the family discovered to their relief that Mr. Seidel was already there. Exhausted from a full evening of printing postcards he had decided to walk home to get some fresh air. They informed the police of the mix-up, thereby ending an exciting night in the sleepy little town. Fortunately, the other nights were uneventful.

The Seidel family's conscientious work resulted in thousands of postcards, and once the printing was complete they went about distributing them to interested businesses. By the time of the August 1939 meeting at the Legion, Joe informed the Post of the venture. He presented a proposition of community service in the "form of distribution of a minimum of 5,000 picture postcards of local, river and mountain scenery…." They were sold to local distributors at a net cost. The minutes recorded, "An expense of $2.00 for Philip Rawson's services was the total expenditures involved, Mr. Sanders' time and efforts were given willingly."[7]

Locals and tourists purchased the crisp black and white cards. Some of the souvenirs stayed unmailed and tucked away in postcard collections while others carried mementos of the Hill Country or town to distant destinations. Eventually the Seidels also had some of the black and white views colored. Though the pastel tones did not represent precisely true images of the scenes, they did offer colorful alternatives to the customer.

With thriving businesses in New Braunfels and a blooming tourism industry, city officials saw the promise of great economic growth as the decade of the Depression drew to a close. Convenient travel clearly played an important role in attracting new enterprises. In August 1939 county commissioners, determined to make more road improvements, voted to surface 16,000 square yards of Hueco Springs Road with asphalt. In September they accepted a bid for $1,366.40. By the beginning of 1940 road laborers were paid an hourly wage of 25 cents. Truck drivers earned 30 cents; a team and teamster received 50 cents.[8]

The Legionnaires, however, were apparently on their own regarding their county road markers. Almost seven years had passed since completion of the project, and many of the signs needed general care and repainting. Through the years Joe had tended to repairs as necessary—sign by sign.

But by 1940 confusion arose over the maintenance of the markers. The Legion minutes of June 10, 1940, recorded:

> A discussion came up in regard to the road signs placed throughout the county by this Post as to the responsibility of upkeep and reconditioning of same, and it was the consensus of the Ex. Board that the county had obligated itself as to the upkeep of same, and that the Post go on record to petition the Honorable County Judge and commissioners to re-condition and repaint these signs.

At the July 8, 1940, meeting the Legion reported that the county judge and commissioners contended that "the County furnish the paint only and that the re-conditioning and painting of the signs were to be done at the expense of the Post."

And so the squabbling went back and forth for a while. In the meantime Joe still spent his share of Saturday and Sunday afternoons relettering road signs at the kitchen table.

With the growing popularity of the New Braunfels area, by April 1941 the Chamber of Commerce asked for a new version of the road map. Joe compiled the updated copy, and it was drawn by a New Braunfels resident, draftsman Roye V. Swartwout. The revised map sported a few changes. Fairy Caverns, which had closed in the mid-1930s, was gone, but a sketch depicted a newly opened attraction in Kendall County, Cave Without A

Guadalupe River above the Third Crossing. (Photo by Joe Sanders.)

Name.[9] Highway No. 66 and Highway No. 2 were now numbered 281 and 81 respectively, and numerous new roads crisscrossed into Guadalupe County to the southeast. Labels and drawings marked Landa Park and New Braunfels' new golf course.[10]

The reverse side of the map outlined detailed statistical information for the town and county. New Braunfels boasted a population of 8,000. The prosperous community advertised two "excellent" hotels, a fire department, ten churches, several bus lines, three summer resorts, and various manufacturing enterprises in flour, feed, textiles, cedar oil, and road materials. Agriculture was a principal county industry consisting of 750 farms averaging seventy-two acres and 375 ranches averaging 800 acres. By this time tourism was earning approximately $500,000 yearly for New Braunfels. The Guadalupe River, referred to as the "Blue Danube of Texas," was among the many attractions suggested to travelers as they drove "along roads lined with wild flowers and ranges abundant with wild deer and other game."[11]

The new road map, out just in time for the summer vacation season, was characterized as "unique among map services for tourists as well as native Texans," showing "every point of interest in the county as well as the

At the end of the Depression the freedom of the open road mirrored a general sense of growth and renewal. Here a driver takes to the country in the Baetge Hill area. (Photo by Joe Sanders.)

corrected distances and how to get there." Portions of Kendall, Blanco, and Hays counties (as well as Bexar and Guadalupe) were again included, and organizations and businesses expected to distribute 10,000 copies of the map that year.[12]

In August 1941 road crews and county commissioners began an ambitious undertaking with the Work Projects Administration (WPA) to hard surface 100 miles of primary county roads.[13] The operation set the tone for the inevitable changes that stood on the horizon. The hard years were over, and it was a time to reap the rewards that patience and perseverance had sown during the Depression.

· · · · Chapter Twelve · · · ·

Is This Trip Necessary? The Joy of the Drive

World War II changed everything. Tourism, suddenly considered an unnecessary luxury, came to a complete stop as the United States mobilized for war. Passenger car production, which had surpassed 3,717,000 vehicles in 1940, ceased in early 1942 as the auto industry converted to military manufacturing. In January 1942 tires were first rationed. Later that year the federal government imposed a national "victory speed" limit of thirty-five miles per hour.[1]

The collective will of the country focused on the goal of winning the war. Every family had someone in the service or knew someone waiting to be shipped overseas to an uncertain future, and the knowledge of the monumental struggle inspired citizens to pull together and make sacrifices at home. On December 1, 1942, authorities enforced gasoline rationing nationwide. "You had to have a ration ticket to get gas at all," Laurie explained. Drivers attached the gas-rationing stamp to the front windshield of a vehicle, and the driver's needs determined the classification of the stamp. Workers who depended on their vehicles in their profession—doctors, for example—received the highest priority. For the average driver, rationing boards allowed only minimal motoring and eventually allotted as little as two gallons per week per car. Everywhere, in store windows and other businesses, signs asking "Is This Trip Necessary?" reminded motorists to conserve and to minimize their travels.[2]

"You weren't supposed to get gas just to go pleasure riding or anything," Laurie recalled. As a mechanic, Joe was fortunate to secure an adequate fuel supply for his needs, because driving was essential to performing mechanical repairs. Indeed, in 1942 he replaced his old Franklin with another vehicle, a late-1930s tan De Soto, but he kept driving to a minimum. "We saved up so that when I wanted to go to Victoria we could get gas to go down there. But otherwise we didn't do much pleasure riding then, just little short trips," Laurie stated.

At a time when new cars were nonexistent and tires were scarce, motorists took special care to maintain their aging vehicles. They had old tires retreaded and coaxed what extra life was left out of strained axles and bearings and clutches. Even so, old autos wore out or were wrecked in accidents, and consequently the number of cars in use nationwide shrank on the average by 4,000 a day.[3]

Road work suffered major setbacks, and many projects stopped. The Texas Highway Department lost a significant part of its workforce as able-bodied men and skilled engineers joined the service. The byways could no longer call out to the explorer, the free spirit, or the family out on a jaunt, but instead the ribbons of pavement summoned caravans of military machines. Ironically, the war did generate its own brand of tourism, a by-product of the mobilization effort. The armed forces stationed personnel at bases and airfields across the country. "It seemed like all the servicemen who were stationed around here came from the North," Laurie commented, "and I understand also that a lot of the local boys were sent way up to the North. And I think the idea around that was to get them acclimated so you knew your whole country."

Some families of servicemen also stayed in the area, and parents and wives of soldiers were willing to pay almost any price to rent rooms from locals to be near their loved ones. Now the Saturday night dance in New Braunfels assumed new meaning as it provided an entertainment outlet for servicemen from the nearby bases, and for a while each week the soldiers could lean on the comforting shoulders of the small town atmosphere. At

Echo Hall, Laurie worked as the hat checker at each dance and spent hours talking to homesick boys. She became a sort of foster mother to sons who were many miles away from home for the first time. Shy soldiers looked for a friendly face or sought encouragement to ask a girl to jitterbug. Laurie Jo, now in high school, always attended the gatherings, and she and her friends exchanged dance partners—perhaps an East Coast Italian or next a Midwest farm boy. Never before had local citizens met so many men from all over the country. And though the Sunday drive had all but stopped, occasionally the Sanders family would take some fellows for a scenic tour. This exposure to the hospitality and beauty of the Hill Country remained vivid in the minds of many servicemen.

Dorothy Kerbow recalled that men from the East, West, North, and South were stationed in the area, and they got acquainted with the natives. "Many of them married girls from here. A boy from Illinois married one of my cousins.... Some of them stayed here. After the war they came back and settled here, and they told their friends...." The migration represented the largest influx of outsiders to move into the area up to that time and foreshadowed coming years of great change and growth.

When the war ended, euphoric tears of victory washed clean the hardships of the past, and the country faced a bright future. The government lifted restrictions on travel, and gasoline rationing ceased on August 15, 1945.[4] As fuel became readily available and new car production resumed, the open road beckoned the liberated wayfarer to zoom on and explore again.

Life returned to normal around Comal County as well, and once again the country rambler and family picnickers became familiar sights along the backroads. With the freedom to drive, Joe seized the opportunity to find and replace worn signboards. On New Year's Day 1946 Laurie recorded in her diary, "J. spent the day with his 'road work'." Other entries followed such as on January 27 when he "spent the day with his road sign business as usual," and on February 17 when he "spent the time with his maps and such."

Sometimes he still devoted long hours to the strenuous sign maintenance. "J. was to leave at 5:30 A.M. for his sign put-upping trip," Laurie wrote for May 6, 1946, "but it was pouring so it was 9:30 [A.M.] when he got off—cool and cloudy. J. got home at 9:30 P.M. and will have to go out again tomorrow."

At this time Joe compiled his records into a book titled "Key To Comal County Road Marking, American Legion Community Service, Comal Post 179." An impressive work, the schedule included a map of Comal County showing each road marker location, exact drawings of every signboard, and diagrams of the crossroads where they were posted—a total of fifty-five intersections and approximately 148 signs. Joe amended markers as routes and mileages changed, and he recorded boards that were replaced or discontinued.[5] He stored his volume in County Judge Clarence Rice's office where it was available for public viewing.[6]

With tourism back into full swing, it was time for a new county road map filled with fresh facts and figures to entice the sojourner. Joe made revisions and issued a 1946 edition with Roye Swartwout. The map touted:

> The hill country north and west of New Braunfels is truly nature's wonderland. The grandeur of the eternal hills, the soft beauty of the green-clad valleys, and rippling waters of the springs and brooks entrance the visitor driving through this scenic country.... Camera fans, artists and lovers of nature revel in these vistas of natural beauty. Every season of the year brings with it a new charm and grace to these hills.... A floral carpet of exotic loveliness is spread over hills, ranges, and valleys in the springtime when millions of bluebonnets, wild verbenas, mountain pinks, wild daisies, mountain laurel, redbud and myriads of other native flowers are in full bloom.[7]

The new map came out in time for the grand centennial celebration of New Braunfels. The community had thrived for over 100 years, and festivities took place on May 10-12, 1946. Joe was a member of the Tours and

"Halloween Deviltry"—for years Joe often repaired and replaced signs damaged by vandalism. In this case, tricksters dismantled markers and rearranged their positions. (Laurie P. Sanders Collection.)

Sightseeing committee, one of the many groups that organized the activities.[8] He took townspeople and visiting VIPs on scenic drives through the countryside, confirming the map's boast that "well marked all-weather roads wind through an ever changing panorama of natural beauty and interesting attractions."

Though the war had temporarily halted most road-building projects, some improvements had continued, and individuals financed work crews with private funding. When Milton Kaderli returned from military service, he was shocked to find Rebecca Creek Road, the road to his family ranch, paved. "Mr. Willard Hill ... owned the other half of this Demijohn Bend," Milton said, "and while I was away in the service he made some sort of a deal with the county.... He put a car-widths pavement all along that road, and it did a lot of good." What had been a primitive path full of mudholes became a smooth clean driving surface.

Road construction really boomed after the restrictions of the war years were lifted. Texas had 26,864 miles of state highways, including 2,557 miles that were not paved in September 1945, but the passage of the Good Roads Amendment the following year assured that state highway funds would be preserved strictly for road purposes, and highway groups heralded the measure as a major success for road development. The highway department initiated improvements that included installing a system of colored signs that reflected the lights of traffic. The state implemented a new Farm-to-Market program and designated many existing and new courses as Farm-to-Market and Ranch-to-Market roads, instigating the growth of a veritable spiderweb of byways across Texas counties. During the first postwar year the highway department built 151 miles of new all-weather thoroughfares for the state farm road system.[9] Crews completed 2,218 miles of farm roads by November 1947, prompting the Texas Good Roads Association to comment in their newsletter "Highway Highlights":

> These 2,218 miles are real roads serving rural areas. They show the tremendous progress being made toward bringing Texas farmers out of the mud. These completed roads involve 300 projects distributed over 172 counties.[10]

After the war Comal County sought to have a number of routes designated as Farm-to-Market roads, which began to crisscross many Hill Country areas. The Old Blanco Road, its segments widened and paved, became a Farm-to-Market road. Familiarly named paths now received impersonal numbers: Bracken Road became 482; Hunter Road became 1102.[11]

While improvements resulted in greater and more efficient travel, the very character of the wild backroads was beginning to change. In the effort to shorten routes and make for safer driving, work crews replaced right-angle turns, huge dips, and winding paths with straighter courses that cut through hills. New powerful highway equipment enabled men to blast and plow and dig through the rugged terrain.[12] Consequently, they abandoned many stretches of roads.

County commissioners had the option to abandon a road. In the late 1940s and into the 1950s the minutes of the Comal County Commissioners Court were riddled with petitions to abandon and close sections of roadways including substantial portions of State Highway No. 66, Smithson's Valley-Spring Branch Road, and Fredericksburg Road. They closed part of Rebecca Creek Road by the old Rebecca Creek School and segments of Old Boerne Road as the course was straightened. They replaced sections of Old Hancock Road, slice by slice, with the "shorter and better route" of Farm Road 306. Perhaps the old Post Road, Highway No. 2, destined to succumb to future Interstate Highway 35, suffered the most change. Constant straightening left stranded fragments of old pavement.[13] Across Texas the highway commission officially removed portions of previously designated state highways, and jurisdiction over them converted to the city or county. Ultimately, the county commissioners court determined the fate of an old state highway. Some of the thoroughfares continued in operation as city and county avenues. Others closed forever, and the land reverted to the

An older crew (Joe, far right) still hard at work posting signs as needed, here at the junction of Hunter Road (FM 1102) and FM 306 near Gruene. (Laurie P. Sanders Collection.)

property owners. The trend toward road abandonment would escalate in the future as the new interstate system would dramatically change road travel across Texas and the nation.[14]

All of the road building, of course, brought to the area greater tourism and more inhabitants eager to explore the Hill Country, and the signposts helped guide a new generation of adventurers seeing every vista for the first time. Surveyor Bill Kolodzie was a new resident who enjoyed the country drive.

> I remember those signs very, very well, because we moved here in '49, and when we moved here, why every weekend we traveled all the backroads in Comal County just to learn them and see what it was, and we followed these signs religiously. A lot of people drove around like that. [They'd] go out to Startzville and "get lost…." It was in the boonies…. They were backwoods….[15]

Joe Sanders continued to repair and post signs as needed, but life for the Sanders family had changed. The death of Mr. Dittlinger in September 1946 meant the end of any chauffeuring duties and the sad loss of a driving companion. By the 1950s, daughter Laurie Jo had graduated from the University of Texas and was starting life on her own.[16] Joe worked long hours at the mill and did not have the spare time he had once had to travel about, but he still issued revised county maps in 1951 and 1955 with artists R. E. Koepp and Phil Rawson, respectively.

Throughout the 1950s a major county issue was the proposed dam and reservoir on the Guadalupe River upstream from New Braunfels and the river drive. The topic caused heated debates among ranchers. Milton Kaderli remembered that a lot of landowners were bitter about it because they didn't want to give up their land. Other county citizens, however, focused on positive aspects of flood control, water conservation, and recreation. Joe personally supported a reservoir, and by the mid-1950s the proposed dam site appeared on the county maps,[17] but the lake would impose inevitable, irreversible effects upon the Guadalupe River Valley and the farmers and ranchers who called it home. The Kaderli homestead on the Demijohn Bend, for example, would witness river backwater miles upstream, flooding trickling rapids and drowning out huge stately cypress trees, leaving an eerie graveyard of ghostly stumps and tangled branches in their place. The lake would represent the most significant change for Comal County. "The lake of course was progress," Kaderli commented, "but as far as the … scenic beauty … along that river … looking back, I would have to say sure it was more beautiful because you had the environment which hadn't been disturbed and the beautiful trees…."

The dam site heralded the end for the villages of Hancock and Cranes Mill and for a way of life known by generations of Hill Country residents. "It never occurred to us that one day there would be no more Hancock Valley and that Hidden Valley would not be [just] a farming area, that that would be more commercial just below the dam," Melba Meckel Roth exclaimed. Gone would be the bluebonnet-strewn meadows along Hancock Road, condemned to be flooded along with stretches of other byways like Cranes Mill Road.[18]

Crews widened Highway 46 to twenty-six feet and added surfaced shoulders in anticipation of heavier traffic. Part of the old course to Smithson's Valley and Spring Branch received its own designation as FM 311. Modern routes like FM 2722 replaced much of Bear Creek Road, bypassing romantic twists and meandering curves and instead cutting straight through cliffs. New courses brought more development and the increasing purchase of land tracts to build subdivisions.[19]

As the roads changed, Joe made the appropriate corrections to his signs. Even as late as 1960 and '61, he ventured out to inspect the signs and replace battered ones. With his little grandson Larry, he would head out in the Hudson, a car he had purchased from Willard Hill in 1951, and they would motor on for a while till he spotted a certain crossroads where a signboard needed straightening or cleaning, or he replaced a weathered and chipped marker with a freshly painted board.[20] But increasingly the county took

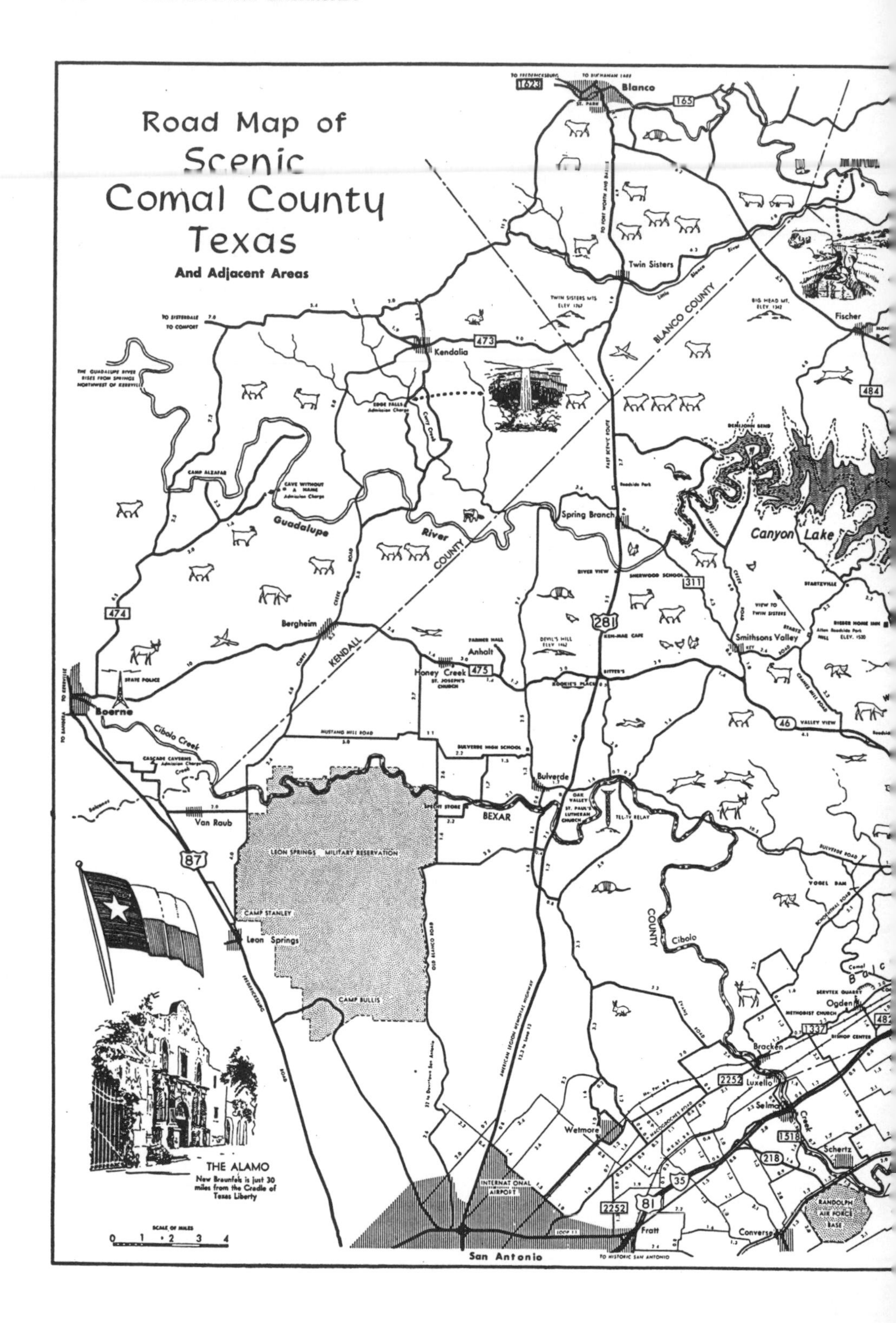
Road Map of
Scenic
Comal County
Texas
And Adjacent Areas
Blanco
1623
165
Twin Sisters
BLANCO COUNTY
Fischer
Kendalia
473
484
Canyon Lake
Spring Branch
Guadalupe
River
COUNTY
311
281
474
Bergheim
KENDALL
Smithsons Valley
Anhalt
Honey Creek
475
Boerne
Cibolo Creek
46
Bulverde
BEXAR
Van Raub
87
LEON SPRINGS MILITARY RESERVATION
CAMP STANLEY
Leon Springs
CAMP BULLIS
COUNTY
Cibolo
Ogden
Bracken
1337
2252
Luxello
Selma
Creek
1518
218
Schertz
Wetmore
35
2252
81
Fratt
Converse
RANDOLPH AIR FORCE BASE
San Antonio
THE ALAMO
New Braunfels is just 30 miles from the Cradle of Texas Liberty
SCALE OF MILES
0 1 2 3 4

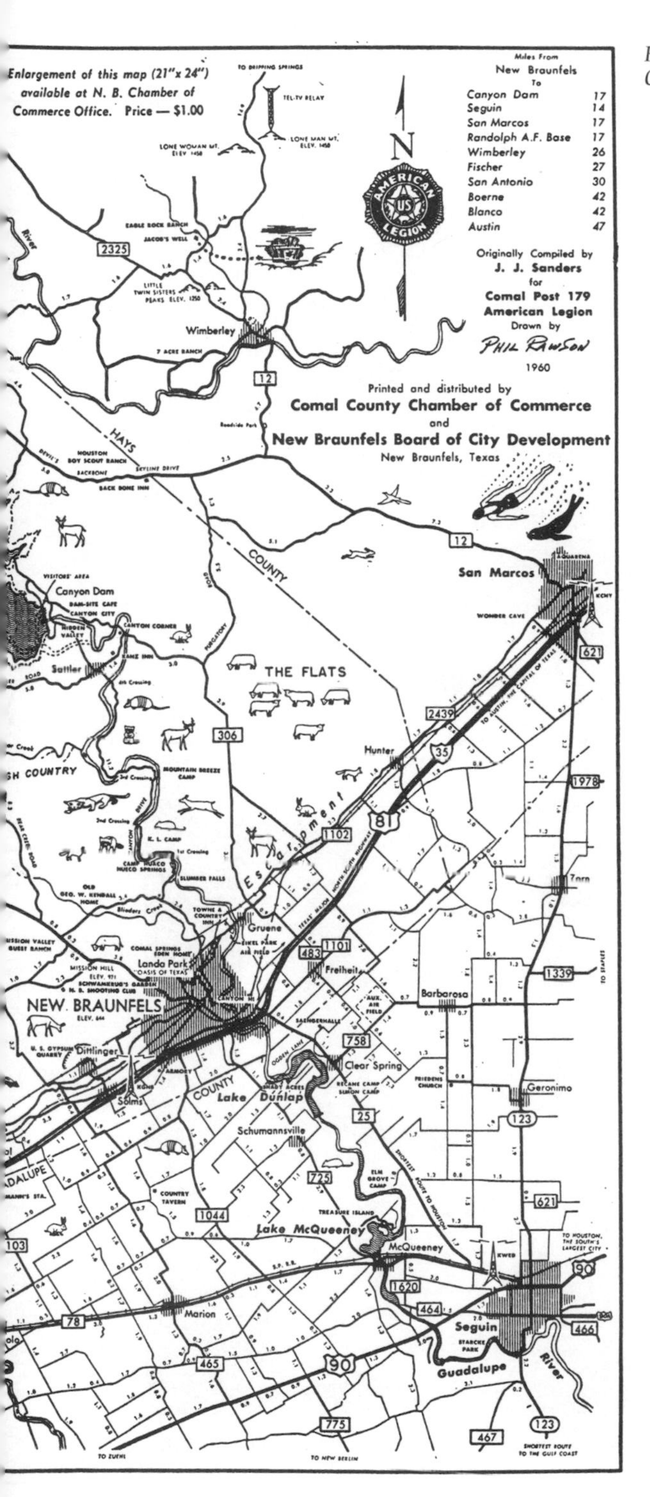

Road Map of Scenic Comal County…1960

over the construction and maintenance of its own markers, and in the years following the impoundment of Canyon Lake in 1964, road crews replaced most of the old signs. From humble beginnings at the Sanders

"Sign pioneer" Joe Sanders as he appeared in the 1963 *San Antonio Express and News* article, "Road Marker Idea Blooms in Comal." (Laurie P. Sanders Collection.)

kitchen table the markers had faithfully guided excursionists for years until crews dismantled them and stacked them in road warehouses. A few workers and souvenir hunters stashed away signs as mementos of the early roads, but mostly crews piled worn out markers and burned them.[21]

In 1960 the state of Texas maintained a total of 57,015 miles in its highway system, and 4,885,300 vehicles were registered. That year the last county road map crediting Joe Sanders came out. New Braunfels had a population of 15,631.[22] By that time many of the private scenic parks had disappeared or were soon to close—unable to compete with commercial recreation facilities and unable to deal with the increased volume of people.[23] In the midst of the growing population and building boom sometimes Joe wished that Comal County could be made into a park. Laurie remembered:

> He said that if he was a billionaire he'd buy the whole county and make it a game preserve ... just keep it natural, because he could see then what was happening. People were coming in mostly from Houston and that section of the country and buying up these ranches up in the hills. A lot of them built homes up there and cleared off the land.... All of that used to be wild territory.

The building development signified the inevitable march of progress—progress that, ironically, Joe had helped bring about through his promotion of tourism. His road map and signs encouraged greater travel and helped open the Hill Country backroads to venturing Sunday drivers. Life was slower paced then, and people were more patient and trusting.[24]

The joy of the drive itself attracted many tourists—when the autocar was still a novel luxury and the dusty roads were wild and woolly. The Sunday tour was a memorable thing, and Joe learned his lessons by losing his way and backtracking and answering the questions of lost motorists such as A. W. Grant so many years past.

An account of Joe's accomplishments appeared in a newspaper feature by county clerk and historian Oscar Haas for the *San Antonio Express and News* in 1963.[25] The story prompted A. W. Grant to write in a Christmas letter to the Sanders family, "I was glad to have been able to inspire it for it was long deserved...."[26]

Haas called Sanders a "sign pioneer" and praised Comal County roads as "among the best marked in the state," as he relayed Sanders' work in his feature:

> All of us some time while motoring on unfamiliar roads, have gotten off onto a wrong road and were going in proverbial "circles" till finally somewhere someone gave correct directions.
>
> For the past three decades, the getting off on wrong roads could not happen here in Comal County, because of a system of road signs developed by J. J. Sanders, resident of New Braunfels.
>
> "When we owned our first car," Sanders said, "we would on weekends enjoy family road-side-lunch picnics along our county's roads. It was at such picnics, when out-of-county motorists stopped to ask road directions, that the idea of a system of road signs was born in my mind."

Joe's life had come full circle. His motto had always been "the only way to do a thing is to do it right." His years of perfectionist deeds to provide a county service had instilled in others the same freedom to explore what he had once experienced when he saw his first car and gleefully drove past the Ohio farms of his childhood. The changes he had seen and helped to bring over the years eventually swept past him as they will always do. But for a time, his work brought excitement and adventure during the youth of the Sunday joyride as people discovered their surroundings. In this context, families shared time together, couples fell in love, and wayfarers ventured around the next bend—throughout the county and across every county in Texas and the nation. The scenic drive for the Sanders family and all families afforded the opportunity to explore, appreciate the beauty of nature, and take time to enjoy life.

AFTERWORD

I've been told you "can't go back," but working with my granddaughter on this book took me back to when I was young, and I find a nostalgic feeling of living those days again, recalling many wonderful memories. So, reading this story was a very happy experience for me. Here in my old age, it brings a touch of my youth.

Each chapter has a different story to tell, but they all tie in together to present a complete picture of the way it was in New Braunfels and Comal County during the era from the 1920s to the 1940s. This book speaks of my husband's dedication to a dream that brought him from the farmlands of Ohio to the Hill Country of Texas. From the time Joe came to New Braunfels, he was very much impressed by the beauty of the surrounding hills, and he set out to fulfill his dream to improve travel and promote the area's scenic attractions. His early fascination with cars and mechanics led to his interest in highways and tourism. He saw the necessity for good roads, and drivers needed signs and maps to show the way. Joe worked diligently to improve driving in the region, not for any gain to himself, but as a service to the community.

I recall all the challenging work that Joe and I did to produce the road signs and the Comal County map, recording all the mileages and directions and every little spot in the road that wanted to be remembered. He met many interesting people as well because he was a person who made friends easily. Many ranchers were more than willing to help and give information that he needed to have for his road mapping and couldn't get except by word of mouth. Joe comes to life in the pages of this book. Any reader will feel they know the kind of person he was. He was loyal to his friends and family and always looking for a way to help.

Joe's motto was "the only way to do a thing is to do it right, otherwise it's a waste of time." He turned road blocks into stepping stones to achieve his goals. I realize that my granddaughter is a perfectionist as was her grandfather. Although she met up with some hardships, in her tireless research she tracked everything—down to the last detail. This volume is a beautiful tribute to the grandfather she never knew, but to me, it is a beautiful tribute to her.

—Laurie P. Sanders

✦ NOTES AND SIDE ROADS ✦

ONE: TAKE TO THE HIGHWAY

1. Laurie Pfau Sanders Interview, July 21, 1988; Joe Sanders, "Places Visited" Travel Diary (author's possession). Joseph John Sanders was born on December 2, 1894, in Ottoville, Ohio. When he was about six months old his father died. His mother went to work and eventually remarried. Joe lived with his grandparents on their farm. In his travel diary in December 1916 Joe reported a population of 600 in Ottoville.

2. *Houston Chronicle*, April 19, 1936, clipping from "Travel Scrapbook," Vertical Files, Center for American History, University of Texas at Austin; Oscar Haas, *History of New Braunfels and Comal County, Texas 1844-1946* (Austin: Hart Graphics, 1968), pp. 69-73, 78-80; Tula Townsend Wyatt Interview, February 18, 1989; *State Highway Department of Texas, Seventh Biennial Report*, September 1, 1928, to August 31, 1930, Austin, Texas, p. 13, Albert B. Alkek Library, Southwest Texas State University, San Marcos; Ron Tyler, ed., *The New Handbook of Texas*, 6 vols. (Austin: Texas State Historical Association, 1996), Vol. 4, "Old San Antonio Road," pp. 1139-1140. Other Spanish routes including trails in Arizona, California, and New Mexico were also named Camino Real.

3. Comal County Commissioners Court Minutes, Vol. A, pp. 1, 5, 7, 15, 21, Comal County Clerk's Office, Comal County, New Braunfels, Texas (hereafter cited as Com. Ct.); Haas, *History of New Braunfels*, pp. 8-9, 17, 22, 69-90; Inventory of County Archives, No. 46, Comal County, WPA Historical Records Survey, February 1937, Part III, "Commissioners' Court," pp. 51-52, copy from the Center for American History, University of Texas at Austin; Larry Kearney Interview, Comal County Road Department, December 5, 1991. Comal County was established in 1846.

4. Com. Ct., Vol. G, pp. 160-161, 36-38, 45, 106.

5. William Kolodzie, Comal County Surveyor, Interview, September 16, 1991 (with the exception of KGNB radio interviews, all tapes and transcripts are in the author's possession); Conversation with Charlotte Boyd, Comal County Clerk's Office, June 1991.

6. Com. Ct., Index to Book H; Com. Ct., Vol. G, p. 45; Road Minutes, Commissioners Court, Comal County, Texas, 1882-1907, 1919, pp. 143-149, 168, 180-187, 192; Kearney Interview.

7. Road Minutes, pp. 5-10, 265; Kearney Interview; Lillian E. Penshorn, "A History of Comal County" (M. A. Thesis, Southwest Texas State Teachers College, San Marcos, 1950), p. 45; Com. Ct., Vol. K, p. 449.

8. Com. Ct., Vol. G, p. 216, Vol. H, p. 592.

9. Albert L. Lewis and Walter A. Musciano, *Automobiles of the World* (New York: Simon and Schuster, 1977), p. 91.

10. "Looking at the Texas Highway System from all angles," pamphlet, San Antonio, Texas: Texas Good Roads Association, ca. 1938, Center for American History, University of Texas at Austin; Lewis and Musciano, *Automobiles,* p. 50; G. N. Georgano, ed., *The Complete Encyclopedia of Motorcars, 1885 to the Present* (New York: E. P. Dutton and Company, 1968, 1973), p. 586; "History of Texas Roads and the Texas Highway Department," Austin: Texas Highway Department, Traffic Services Division, ca. 1948, pp. 2-3, Center for American History, University of Texas at Austin. The auto race was held in Rhode Island on September 7, 1896.

11. Com. Ct., Vol. I, pp. 240, 325-326.

12. Harry Landa, *As I Remember...* (San Antonio: Carleton Printing Company, 1945), pp. 79-80. Landa recounted a humorous motoring story about driving home from San Antonio one evening with his nephew and a chauffeur:

    One day we came to San Antonio to spend the day, and started home about dusk. The crossing on the Cibollo [creek] was very bad, rough and rocky, and the hill on the Comal County side was very steep and hard to climb. The automobile stalled as it started up the hill, so, thinking it would lighten the load, Harry [the nephew] and I jumped out and the car went on. Well, here I was with a sick boy on a lonesome country road and sixteen miles from home.

    I thought surely the chauffeur would soon see that we were not in the car and return for us, but that hope was in vain. When he reached New Braunfels, he drove the car into the garage, and called it a day.

    Landa and his nephew walked twelve miles before finally reaching a friend's house in the village of Solms, four miles outside of New Braunfels. The friend drove them home.

13. Milton Kaderli Interview, March 11, 1991. Both the Pfeuffer Store and Henne's were located downtown on West San Antonio Street. The Pfeuffer Store was on the corner of San Antonio and Castell streets.

14. *The Inland Merchant,* "Younger Every Year," March, 1924, pp. 27-28, magazine excerpt located in "Somers V. Pfeuffer" Vertical File, Center for American History, University of Texas at Austin.

15. "Texas Good Roads Association," pamphlet, n.p.: Texas Good Roads Association, n.d.; Lewis and Musciano, *Automobiles,* p. 140; "History of Texas Roads and the Texas Highway Department," pp. 2-3.

16. A runabout was a smaller open sporty vehicle, usually with two seats. A touring car was also open but larger and seated four or more people.

17. Lewis and Musciano, *Automobiles,* p. 190; Com. Ct., Vol. K, p. 360.

18. John B. Ruppel Interview, conducted by Herb Skoog (with Fred Oheim) for "Reflections," KGNB radio program, New Braunfels, Texas, aired June 5, 1977 (KGNB tapes and transcripts for "Reflections" are housed at the Sophienburg Archives, New Braunfels, Texas); Kearney Interview.

19. Floyd Clymer, *Henry's Wonderful Model T, 1908-1927* (New York: Bonanza Books, 1955), pp. 32-35; Ruppel Interview; A. L. Dyke, *Dyke's Automobile and Gasoline Engine Encyclopedia*, Twelfth Edition (St. Louis, Missouri: A. L. Dyke, Publisher, 1920), pp. 430-438.

20. Johnny Ruppel later owned his own dealership, Ruppel Auto Company, on South Seguin Street.

21. Com. Ct., Vol. K, pp. 614, 633, 344, 586, 593.

22. *New Braunfels Herald*, October 23, 1914.

23. Com. Ct., Vol. L, pp. 54, 209, 214.

24. Joe Sanders, "Places Visited" Travel Diary; Sanders Interview, July 21, 1988. Sanders worked at the Perfection Biscuit Company dipping cookies in chocolate. When he worked at General Electric he received his pay in gold coins each Saturday.

TWO: NEW CARS, OPEN ROADS, AND A YOUNG ADVENTURER

1. Sanders Interview, July 21, 1988.

2. Hip Mengden Interview, August 9, 1991. Dittlinger always went by his first initial, "H," rarely ever referring to his full first name or even signing it.

3. Dorothy Wimberley Kerbow Interview, February 14, 1989; Sanders Interview, August 25, 1988; Mengden Interview; Kearney Interview.

4. "Looking at the Texas Highway System from all angles"; Lewis and Musciano, *Automobiles*, p. 258; Frank Lively, ed., *Texas Highways*, Fiftieth Anniversary Edition, Austin: Texas Highway Department, Travel and Information Division, Vol 14, No. 9 (September 1967), p. 32; *State Highway Department of Texas, Seventh Biennial Report*, pp. 93, 97; Com. Ct., Vol. L, pp. 447-448, 452-454, 562-563. The route of Post Road (and present I. H. 35) to San Antonio, combined with the layout of the counties, meant that drivers left Comal County and actually drove through a small strip of Guadalupe County before reaching Bexar County at Cibolo Creek. Scarification means the breaking up and loosening of rock and rubble.

5. Sanders Interview, July 21, 1988; United States War Department, Certificate for Joe Sanders, Government Request For Transportation, January 1919 (author's possession).

6. Sanders Interview, July 21, 1988; Mendgen Interview; Franziska Dittlinger Liebscher and Maria Liebscher Interview, February 19, 1990.

7. Lewis and Musciano, *Automobiles*, p. 261.

8. Com. Ct., Vol. M, p. 1. A few months later the court raised the salary to $150 a month (Vol. M, p. 42).

9. Fred Oheim, comment taken from Alfred Liebscher Interview, conducted by Herb Skoog (with Fred Oheim) for "Reflections" KGNB radio program, New Braunfels, Texas, aired August 7, 1977; Kerbow Interview; Com. Ct., Vol. M,

p. 400. A good example of some of the dangers of early driving appears in an entry in the commissioners court minutes that mentions the need for an iron or concrete railing at "Baese's Place on Post Road." During a recent rainstorm, a car had slipped off the road into a watery ditch and four people drowned.

10. *San Marcos Record*, January 9, 1920; A. Liebscher Interview; Mengden Interview. Cibolo Creek was the boundary line between Comal and Bexar and Bexar and Guadalupe counties.

11. T. H. Webb, compiler, Assistant State Highway Engineer, *Texas State Highway Commission: Maintenance Manual Inaugurating the Patrol System of Maintenance* (Austin: Texas State Highway Commission, Published for the Information of County Commissioners, County Road Superintendents and Patrolmen, 1922), p. 7; *State Highway Department, Seventh Biennial Report*, p. 25; *General Laws of the State of Texas Passed at the Second Session of the Fourteenth Legislature* (Houston: A. C. Gray, State Printer, 1875), Chapter 108, pp. 167-169; "State Department of Highways and Public Transportation, Introduction," booklet, ca. 1986, copy from "Roads" Vertical Files, Center for American History, University of Texas at Austin. First-class roads also were not to exceed a maximum width of sixty feet.

12. Com. Ct., Vol. M, pp. 161, 185-186, 276, 278, 345, 347, 482.

13. *Texas State Highway Commission Maintenance Manual...* p. 7; Com. Ct., Vol. M, p. 115; Com. Ct., Vol. N, p. 91; *State Highway Department of Texas, Seventh Biennial Report*, p. 25.

14. *Automobile Trade Journal*, New York Silver Jubilee Show Number, Philadelphia: Chilton Company, Vol. 29, No. 7 (January 1, 1925), p. 111; Alamo Classic Car Showcase and Museum, New Braunfels, Texas, Franklin display; Lewis and Musciano, *Automobiles*, p. 281; Sanders Interviews, July 21, 1988; August 17, 1990; Mengden Interview.

15. Tad Burness, *Cars of the Early Twenties* (Philadelphia: Chilton Book Company, 1968), pp. 32, 112-117; Jerold L. Kellman, ed., *Cars of the 30s*, Skokie, Illinois: published by *Consumer Guide* magazine, Classic Car Bi-Monthly, Vol. 269 (May 1980), pp. 43-44; Dyke, *Automobile and Gasoline Engine*, p. 189. Lindbergh's "Spirit of St. Louis" plane had an air-cooled engine.

16. Clymer, *Henry's Wonderful Model T*, p. 122, inside cover jacket summary; Lewis and Musciano, *Automobiles*, p. 259; Burness, *Cars*, p. 244; Com. Ct., Vol. M, p. 256.

17. Sanders Interview, June 30, 1991; Kerbow Interview; Ruppel Interview; Burness, *Cars*, pp. 38-39. According to auto books of the time there were three possible elements employed in car-heating methods—the use of hot water, exhaust gas, and hot air (Dyke, *Automobile and Gasoline Engine*, p. 192).

18. Clymer, *Henry's Wonderful Model T*, inside jacket and pp. 16, 191; Ruppel Interview; A. Hyatt Verrill, *How To Operate A Motor Car* (Philadelphia: David McKay, Publisher, 1918), pp. 102-104; H. Clifford Brokaw and Charles A. Starr, *Putnam's Automobile Handbook* (New York: G. P. Putnam's Sons, The Knickerbocker Press, 1918), p. 32.

19. Sanders Interview, June 30, 1991; Clymer, *Henry's Wonderful Model T*, p. 191.

20. Jarvis Gregory Hillje Interview, conducted by Herb Skoog for "Reflections" KGNB radio program, New Braunfels, Texas, aired April 22, 1979. The specific destination of the friend's trip was Medina Dam outside San Antonio.

21. To find a small leak, it was helpful to submerge the tube in water and watch for the air bubbles (Dyke, *Automobile and Gasoline Engine*, p. 568).

22. Clymer, *Henry's Wonderful Model T*, p. 19; Mengden Interview; Dyke, *Automobile and Gasoline Engine*, p. 775.

23. Egon and Ella Jarisch, Interview, conducted by Herb Skoog for "Reflections" KGNB radio program, New Braunfels, Texas, aired April 27, 1980.

24. Joe Sanders, "Places Visited" Travel Diary. The cord tire, as opposed to the fabric tire which used closely interwoven fabric for the carcass, was composed of loosely woven cords or threads. The Goodyear cord was a multiple-cord tire made up of six to eight layers of small cords or threads (Dyke, *Henry's Wonderful Model T*, p. 559).

25. Sanders Interview, August 13, 1990. Mark Howell's *Racing Stutz* (New York: Ballantine Books, 1972) provides an interesting history of the Stutz cars.

THREE: SWEET SUNDAY DRIVES

1. Walton Hoffmann Interview, conducted by Herb Skoog for "Reflections" KGNB/KNBT radio program, New Braunfels, Texas, aired March 22, 1987. The New Braunfels chapter of the Legion was organized about 1920.

2. Sanders Interview, July 21, 1988; Erwin Henk Interview, February 21, 1991.

3. Sanders Interviews, Fall 1992, Summer 1990, January 23, 1993; Laurie P. Sanders and Ralph Pfau Interview, January 8, 1989. Laurie Pfau Sanders was born on September 30, 1902. For more information on Sarita, Texas, see: Laurie E. Jasinski, "Memories of Early Sarita," *The Journal of South Texas*, South Texas Historical Association, Vol. 6, No. 1 (1993), pp. 24-53. William Gano ("Uncle Billy") was a well driller in Sarita.

4. Sanders Interviews, August 27-28, 1992, June 30, 1991; Lewis and Musciano, *Automobiles*, pp. 237, 296; Peter Roberts, *A Pictorial History of the Automobile* (New York: Grosset and Dunlap, 1977), p. 150. Though electric vehicles had some enthusiasts early in the twentieth century, production of most had stopped by the mid-1920s.

5. Sanders Interviews, July 21, 1988, June 30, 1991; Georgano, *Encyclopedia*, p. 652; Clymer, *Henry's Wonderful Model T*, p.162. The Stoddard-Dayton was manufactured in Dayton, Ohio, from 1904 to 1913. From 1913 to 1924 black was the only color offered on a Model T.

6. Laurie P. Sanders Scrapbook, miscellaneous (undated) newspaper clippings (author's possession). The *Victoria Daily Advocate* commented on the hitchhiking adventure:

   Two manly young college chaps, Harvey Fite and Gene Cullum, left Houston last Sunday morning, on foot, en route to Medina Lake, which is above San

Antonio, via Victoria.... Victoria is scheduled to be their first stop. It is the intention of the boys to walk the entire distance unless they can get a ride on some kind of vehicle of transportation other than trains.... It is not known when the young men will arrive in Victoria, however, if they are fortunate enough to catch some motorist who is coming through to this city from Houston, they may arrive at any time.... The boys are planning to make the return trip to Houston in the same manner that they are traveling at present.

A later clipping stated that they arrived in Victoria the following Tuesday adding: "The young men stated that they had a very pleasant journey and that they only had to walk about twenty miles, passing motorists picking them up and giving them a lift."

According to Laurie Sanders, the term "hitchhiking" was not used in the early 1920s. Drivers were not accustomed to seeing travelers along the road. This changed during the Depression, however, when many people searching for work traveled across the country.

7. The United States census for 1920 recorded a population of 3,590 for New Braunfels (*The Texas Almanac and State Industrial Guide*, 1925 [Dallas: A. H. Belo, 1925], p. 63).

8. Sanders Interviews, July 21, 1988, August 9, 1990; *San Antonio Express*, August 21, 1938.

9. *San Antonio Express*, October 4, 1925; Sanders Interview, August 25, 1988; *New Braunfels Herald*, April 10, 1925, July 24, 1925, October 2, 1925, November 5, 1926. *Ausländer* is German for foreigner or alien.

10. Kerbow Interview.

11. Sanders Interview, August 9, 1990; Kaderli Interview.

12. *Austin American*, April 3, 1924. Advertisement reads: "Red Ball Bus cars leave daily for San Antonio every hour on the half from 5:30 A.M. to 6:30 P.M. Buick sedans leave 7:30 A.M., 8:30 A.M., 1:30 P.M., 2:30 P.M., 5:30 P.M. Stations: Maverick Cafe, Rogers Cafe. Phone 6564."

13. Mengden Interview; Kerbow Interview; Jim Owens Telephone Interview, August 4, 1991, San Angelo county agent. Owens was an employee with bus companies for fifty years.

14. Lively, *Texas Highways*, September, 1967, p. 32; *San Antonio Express and News*, September 26, 1965; *Texas Almanac*... 1936, p. 330; *State Highway Department of Texas, Seventh Biennial Report*, p. 97.

FOUR: CATTLE GUARDS AND BUMPER GATES

1. Brokaw and Starr, *Handbook*, pp. 26–32, 244–250; Verrill, *How to Operate a Motor Car*, pp. 102–104; Dyke, *Automobile and Gasoline Engine*, pp. 511–520.

2. Verrill, *How to Operate a Motor Car*, p. 100.

3. Brokaw and Starr, *Handbook*, p. 244.

4. Sanders Interview, August 25, 1988. In the 1920s the road now known as Fredericksburg Road went uphill to the present location of Schwamkrug's Steakhouse and then turned west to meet Highway 46. At that time and for many years, Walnut Street stopped in town and did not extend up the hill west from New Braunfels to join Highway 46.

5. Sanders Interview, January 1993; Clymer, *Henry's Wonderful Model T*, Introduction; Lewis and Musciano, *Automobiles*, p. 299.

6. As a toddler, Laurie Jo Sanders referred to herself as "You." Hence she really meant, "I don't want to go to the countryside!"

7. Com. Ct., Vol. K, pp. 156, 174, 225, 241. The Austin Brothers Company of Dallas built the Gruene bridge for $5,680 in 1910. Shortly afterwards the court ordered a survey for a new road heading east from the crossing and up Gruene Hill. In April 1914, the court ordered a new county road from the New Braunfels city limits to Gruene Crossing at a cost of $2,325 ending near the "corner of the rock fence near the Concrete Bridge." A new road from Gruene's Hill to Blanco Road and the MKT railroad was also ordered (Com. Ct., Vol. K, p. 612).

8. Sanders Interview, January 1993; Kearney Interview; J. C. Riley Interview, February 17, 1989. Austin Hill is located roughly at the present location of the Conrads Road and the Kohlenberg Road Exit 193 on I.H. 35 several miles northeast of New Braunfels.

   Present Hunter Road (FM 1102) follows much the same route as the Post Road. J. C. Riley opened Riley's Bar in Hunter in 1933 and owned and operated the establishment for over fifty-five years. The tavern was located exactly halfway between Austin and San Antonio on the Post Road.

9. Sanders Interview, August 9, 1990; Kearney Interview; Kolodzie Interview. Present FM 2722 roughly follows the old Bear Creek Road route. Three parts of the original course still remain, branching off FM 2722, and are called Old Bear Creek Road No. 1, Old Bear Creek Road No. 2, and Old Bear Creek Road No. 3.

10. Mengden Interview. Hip Mengden was the son of Hippolyt Mengden and Amalie Dittlinger Mengden.

11. Laurie Sanders Jasinski Interview, August 21, 1991.

12. *San Antonio Express and News*, September 26, 1965. The road log for the 1928 tour was reprinted in the 1965 newspaper issue.

13. Sanders Interview, January 1993; Mengden Interview; *Automobile Trade Journal*, p. 200; Burness, *Cars*, pp. 40–41.

14. Sanders, Miscellaneous Interviews; William Barnett Interview, March 15, 1988; Mengden Interview; Kolodzie Interview. William Barnett worked on the Nordan Ranch (Seven Eleven Ranches) located in the Hill Country in the area near Cranes Mill and the Guadalupe River (near present Canyon Lake).

15. The author heard descriptions of at least two types of bumper gates. One style had a pole in the center, as explained in the text. Another type of gate, attached to a pole at the fence, could swing open in either direction.

16. Sanders Interview, October 1991.

FIVE: SIGNS OF THE TIMES: TRAVELERS, BUMS, ENTREPRENEURS

1. *San Antonio Express*, October 29, 1929.
2. Ibid. Baltimore County in Maryland came in second, and Los Angeles County in California came in third. Ironically, the *San Antonio Express* on October 29, 1929, the same day as the great stock market crash, devoted a massive issue to business and industry in the area. The edition also printed a history of the Southland Greyhound bus line and its predecessor, the Red Ball.
3. Sanders Interviews, July 21, 1988, August 25, 1988.
4. Sanders Interview, Summer 1993; American Legion Minutes, Comal Post 179, September 30, 1929; *New Braunfels Herald-Zeitung*, March 28, 1996.
5. Sanders Interview, July 21, 1988. San Antonio city directories listed Albert W. Grant as managing editor of the Express Publishing Company from 1917 into the 1940s.
6. Laurie Sanders Jasinski and Laurie P. Sanders Interview, August 21, 1991.
7. Sanders Interview, August 14, 1990; Iris Timmermann Schumann, "The Great Depression, 1929-1939: A Period of Changing Social Attitudes Toward Community Self-Reliance in New Braunfels and Comal County, Texas" (M. A. Thesis, Southwest Texas State University, San Marcos, 1980), p. 27.
8. Schumann, "The Great Depression," pp. 30-32; Sanders Interviews, August 23, 1990, July 1993.
9. Sanders Interview, August 14, 1990; Jasinski and Sanders Interview, August 21, 1991. The word "bum" was a common (somewhat derogatory) term during the Depression and was often used in a broad sense to describe wandering, homeless, unemployed men.
10. Kaderli Interview. Automobile dealers would even take eggs in trade.
11. Mengden Interview, August 9, 1991. Actually, the town of Flatonia was named for businessman F. W. Flato (Tyler, ed., *The New Handbook of Texas*, Vol. 2, p. 1027).
12. *Texas Highways*, pp. 141-144; miscellaneous newspaper clippings from November 1934, Texas Highway Department clippings, Vertical Files, Center for American History, University of Texas at Austin.
13. *San Antonio Express*, April 26, 1931.
14. At that time U. R. Hellmann served as post commander.
15. *New Braunfels Herald*, March 27, 1931. An article about the contest appeared in the March 20, 1931, issue of the *Herald*:

> New Braunfels has never had an official slogan, although it has been known as "The Beauty Spot of Texas" for several years. It is the desire of the Lions to assist the city in securing an official, permanent, appropriate slogan; hence the decision to sponsor the contest.

16. Ibid., May 8, 1931.
17. Sanders Interview, August 14, 1990; Schumann, "The Great Depression," p. 25. Joe was a member of the City Board of Development (sometimes called the Board of City Development). It was an active civic organization, but Laurie Sanders claimed that the group did a lot of arguing, and she referred to it as the Board of City Devilment.
18. Sanders Interview, August 14, 1990. Like most people, Joe worked on Saturdays, so the family did not go driving then. The Sunday drive was, literally, on Sunday. People worked on Saturdays and stores stayed open until 10 P.M. The practice of working on Saturdays changed during the 1950s (per Sanders Interview, October 31, 1996).
19. Kolodzie Interview; Kearney Interview; author's observations of early maps at Comal County Road Department, New Braunfels, and at the Center for American History, University of Texas at Austin.
20. *Texas Highways*, p. 130.
21. United States Geological Survey Maps of Comal County, Texas, including New Braunfels, 1920-21; New Braunfels, Texas, 1927—unsurveyed edition west of New Braunfels; and Smithson's Valley Map, April 1927.
22. Mengden Interview; Sanders Interview, July 21, 1988.
23. Jasinski and Sanders Interview, August 21, 1991; Sanders Interview, July 1993; Alamo Classic Car Showcase and Museum, New Braunfels, Texas; Lewis and Musciano, *Automobiles*, pp. 366-367; Kellman, ed., *Cars of the 30s* (May 1980), p. 39. Balloon tires first came out about 1926.
24. Sanders Interview, July 1993; Burness, *Cars*, p. 112.
25. Franziska Dittlinger Liebscher and Maria Liebscher Interview.

SIX: SHOWING YOU THE WAY

1. Sanders Interviews, Winter 1989, October 31, 1991, October 1991, Winter 1994; Schumann, "The Great Depression," p. 74; Kearney Interview. The route of Highway No. 2 came into New Braunfels from the northeast and crossed the rickety iron Guadalupe Bridge (by the textile mill). State Highway 66 is not to be confused with the nationally famous U.S. 66 (Route 66), which passes through Amarillo in northern Texas.
2. *Texas Highways*, pp. 146-151.
3. *Texas State Highway Commission: Maintenance Manual* ... p. 37.
4. Sanders Interviews, January 1989, July 21, 1988. During this time Joe continued to plan scenic excursions for civic groups including the Business and Professional Women's Club (Joe Sanders Legion Scrapbook, Mayme Buske, New Braunfels, Texas, letter to Joe Sanders, New Braunfels, Texas, undated—approximately April 1932, author's possession). A letter from the club expressed

"thanks and appreciation to you [for] planning and taking charge of the scenic drive on Saturday, April 9th, during our District Conference."

5. Miscellaneous Clippings, Joe Sanders Legion Scrapbook (author's possession); Legion minutes, Comal Post 179, New Braunfels, Texas, 1932-33; Wally Stahl Interview, February 28, 1991; Walton F. Hoffmann, compiler, "Birth of the American Legion In New Braunfels, Texas," Sophienburg Archives, New Braunfels, Texas; Hoffmann Interview. Comal Post 179 finally constructed a building of its own in 1949.

6. Legion minutes, March 20, 1933. The author has also seen the spelling of George Eikel Schmidt (without the hyphen).

7. Sanders Interviews, July 21, 1988, August 14, 1990; Legion Scrapbook; Clarence Rice Telephone Interview, February 21, 1989. Rice was a Comal County judge from 1947 to 1966.

8. J. J. Sanders, "Key to Comal County Road Marking American Legion Community Service Comal Post 179," Compiled By J. J. Sanders for Comal County, 1946, Comal County Road Department, New Braunfels, Texas. Legionnaires had to get Saturdays off in order to do the signposting.

9. Both sets of markers in Hays County were located on or near Purgatory Road: one set at the junction of Purgatory Road and what is now Hays County Road 214 (also called Hugo Road) and the other at the junction of Purgatory Road and Ranch Road 32. (Note that the Hays County section of Purgatory Road is designated Hays County Road 215.)

10. Legion Scrapbook, undated clipping.

11. *The American Legion Monthly*, Vol. 14, No. 3 (March 1933), p. 32.

12. Emil O. Haas had been commander of Comal Post 179 in 1931 and 1932 and later served on the Post's executive board. He owned the A Cafe.

13. *Road Map of Comal County Texas*, Comal Legion Post 179, Compiled by J. J. Sanders, Drawn by C. H. Bernstein, 1933, Center for American History, University of Texas at Austin.

14. Legion Scrapbook, undated clipping.

15. *New Braunfels Herald*, June 16, 1933.

16. Legion Scrapbook, undated clipping.

17. W. G. Jones, Secretary-Manager, Motor League of South Texas, Houston, Texas, to J. J. Sanders, New Braunfels, Texas, July 22, 1933 (author's possession). State Highway No. 9 ran from Corpus Christi through San Antonio and the Texas Hill Country, on through San Angelo and north to the Texas Panhandle. State Highway No. 20 ran from the area of the southeastern Texas counties of Austin and Washington, to the city of Austin, then northwest towards Llano.

18. *San Antonio Express*, July 23, 1933.

19. *San Antonio Express and News*, February 16, 1963. A 1933 blueprint copy of the *Road Map of Comal County* had the slogan "Showing You The Way" printed on it.

SEVEN: CELEBRATION AND CONTROVERSY: THE ROAD TO THE CENTENNIAL

1. Com. Ct., Vol. H, p. 322. In the 1990s one of the last remaining limestone mile markers in Comal County stood on Fredericksburg Road on the edge of Landa Park. It was designated a historical landmark by the New Braunfels city council on February 14, 1983. According to its commemorative plaque, the stone mile markers were erected in November 1854 and used to measure the distances from New Braunfels to Fredericksburg along the original county road system.

2. *New Braunfels Herald*, June 23, 1933.

3. "History of Texas Roads and the Texas Highway Department," p. 16; *Texas Highways*, pp. 24, 155, 160; Joe O. Naylor, *Naylor's Epic-Century Magazine: The Illustrated Historical Quarterly of Texas*, Vol. 3, No. 1 (April 1936), pp. 5-9, 64; *Dallas Morning News*, May 24, 1936.

4. *New Braunfels Herald*, July 14, 1933, July 28, 1933, August 25, 1933.

5. Ibid., August 25, 1933.

6. Ibid., July 14, 1933, December 8, 1933. The route of State Highway 66 went through western Comal County on to Blanco and Johnson City and followed a course similar to the current U.S. 281 (Kearney Interview).

7. Schumann, "The Great Depression," pp. 50-54, 75-76, 83-84; *Family Encyclopedia of American History* (Pleasantville, New York: The Reader's Digest Association, 1975), "Civil Works Administration," p. 227; "Works Progress Administration," p. 1261; *New Braunfels Herald*, December 1, 1933, December 8, 1933.

8. *New Braunfels Herald*, December 29, 1933. Charles W. Scruggs was publisher of the *Herald* from 1930 to 1957.

9. Joe Sanders Legion Scrapbook, newspaper clipping, ca. May 1934.

10. Schumann, "The Great Depression," Appendix VI, p. 145.

11. *New Braunfels Herald*, May 18, 1934.

12. Ibid., May 25, 1934, December 29, 1933; Legion Scrapbook (author's possession).

13. Schumann, "The Great Depression," pp. 77-78; *New Braunfels Herald*, July 13, 1934, February 1, 1935, April 10, 1936; John R. Fuchs, *A Husband's Tribute To His Wife* (San Antonio: The Naylor Company, 1938), pp. 144-149.

14. *Texas Highways*, pp. 157, 160; Frank Maddox, "Know the Roads You Ride On," *Texas Parade*, July 1936, pp. 7, 27.

15. Charles E. Simons, "Traveler's Oasis: The Story of Texas' Roadside Parks," *Texas Parade*, September 1936, pp. 3-4.

16. *Texas Highways*, p. 157.

17. Simons, "Traveler's Oasis," pp. 5, 24; *Family Encyclopedia of American History*, "National Youth Administration," p. 777. By 1996 the Texas Department of Transportation planted 60,000 pounds of wildflower seeds annually along

Texas highways ("Pocket Facts," Austin: Texas Department of Transportation, Winter 1997).

18. *Dallas Morning News*, May 24, 1936.

19. *New Braunfels Herald*, July 13, 1934.

20. Legion minutes, April 6, 1936; Legion Scrapbook, newspaper clipping (September 12, 1934). The clipping, titled "Road Beautification Prizes Awarded," reported, "In the filling station division J. J. Sanders reported Demuth station No. 3 took first place, Friendly Home station, second, and Jonas station, third."

21. Jasinski Interview, August 21, 1991.

22. *New Braunfels Herald*, April 19, 1935.

23. Sanders Interviews, August 14, 1990, Spring 1994; Burness, *Cars*, pp. 5, 32, 117.

24. Sanders Interview, July 21, 1988. Readers may recall that in the 1950s while planning Disneyland, Walt Disney referred to his designers as "Imagineers" (Richard Schickel, *The Disney Version: The Life, Times, Art and Commerce of Walt Disney* [New York: Simon and Schuster, 1968], p. 310).

25. *The American Legion Scenic Road Map of Comal County Texas*, Comal Post 179, New Braunfels, Compiled by J. J. Sanders, Drawn by P. N. Rawson, Centennial Edition, 1936, Sophienburg Archives, New Braunfels, Texas.

26. "Looking at the Texas Highway System from all angles"; *Texas Almanac* ... 1936, p. 328; *Eleventh Biennial Report of the State Highway Department*, September 1, 1936 to August 31, 1938 (Austin: Texas Highway Department, 1938), p. 14. In summer 1936 New Braunfels boasted the return of another beautiful attraction—Landa Park—which had closed after its owner, an investment company, suffered financial losses due to the Depression. Civic organizations campaigned for the city to purchase the park, and citizens voted overwhelmingly to buy the land. On May 30, 1936, over 1,000 people came out to clean Landa Park, which had become overgrown with weeds and brush. Among the workers were Joe Sanders and a large Legion contingent (E. P. "Pete" Nowotny Interview, June 24, 1991); Sanders Interview, Winter 1989; *New Braunfels Herald*, December 31, 1964.

EIGHT: THE GUADALUPE RIVER DRIVE

1. *San Antonio Express*, July 23, 1933.

2. Com. Ct., Vol. I, pp. 26-30; Ibid., Vol. H, pp. 535-536.

3. Ibid., Vol. L, pp. 435, 484. On the river road, crossings were referred to by number (first, second, etc.). Spiess Bottom was probably named for Hermann Spiess, an early settler and the last manager of the German Emigration Society. In 1849 he bought land along the river, including the Waco (Hueco) Springs area.

4. Ibid., Vol. N, p. 91; *New Braunfels Herald*, March 22, 1935.

5. *San Antonio Express*, July 23, 1933.

6. Sanders Interviews, July 21, 1988, August 9, 1990.
7. Melba Meckel Roth Interview, conducted by Herb Skoog for "Reflections," KGNB/KNBT radio program, New Braunfels, aired November 13, 1983.
8. Sanders Interview, March 6, 1995.
9. *San Antonio Express,* April 26, 1931; *New Braunfels Zeitung-Chronicle,* April 12, 1964. The same Philip N. Rawson who worked as artist on the county map leased Waco Springs Park. There are several variations of the Waco spelling, including Hueco and Huaco. Waco Indians reportedly camped on the Guadalupe River, and the springs were most likely named for them. The term "Waco" was also an Indian word meaning "empty," in regard to the "spring's habit of suddenly going dry for no apparent reason." The term "hueco" is a common name for the "prickly little sand-burs which infest the neighborhood of the springs." In the 1990s the spelling still varied. There was a Huaco Springs Campground and Hueco Springs Loop Road.
10. *San Antonio Express,* April 26, 1931.
11. "Waco Springs Park," brochure, ca. early 1930s, Center for American History, University of Texas at Austin.
12. *San Antonio Express,* July 23, 1933.
13. Kaderli Interview.
14. Sanders Interview, August 25, 1988.
15. *San Antonio Express,* November 1, 1936, July 23, 1933; Sanders Interview, August 9, 1990; Mengden Interview; Kaderli Interview.
16. Sanders Interview, August 9, 1990; March 6, 1995.
17. Mengden Interview; Jasinski Interview, August 21, 1991. Slumber Falls Camp opened in the 1930s. Another tourist camp, Mountain Breeze Camp, located farther down the Guadalupe River Drive, opened after World War II.
18. *San Antonio Express,* July 23, 1933.
19. Com. Ct., Vol. O, pp. 46, 399, 587, 595.
20. Roth Interview. Once workers started to pour concrete for a bridge they could not stop until completion, because fresh concrete will not bond to cured (dried) concrete. "Ready mix" is a general term used to describe packaged concrete that has all the proper proportions of cement and aggregates (sand and gravel) already mixed in. Good quality concrete has specific proportions of cement, fine and coarse aggregates, entrained air, and water (Richard Day, *The Practical Handbook of Concrete and Masonry* [New York: Fawcett Publications, Inc., 1969], pp. 4-17, 32-25; phone calls to American Concrete Institute, Detroit Michigan, and Portland Cement Association, Skokie, Illinois, 1997).
21. Fred Oheim Interview, conducted by Herb Skoog for "Reflections," KGNB radio program, New Braunfels, Texas, aired December 12, 1976; Jasinski Interview; Sanders Interview, August 9, 1990.

NINE: THE SKYLINE DRIVES

1. *San Antonio Express*, July 23, 1933.

2. Laurie Sanders Jasinski (with Laurie P. Sanders) Interview, August 9, 1990.

3. Roth Interview; Kearney Interview; Kerbow Interview; *San Antonio Express and News*, February 16, 1963; Com. Ct., Vol. E., p. 281; Sanders Interview, August 9, 1990.

4. Roth Interview. Solid rubber tires were the first kind of tires on early vehicles. For many years they were still used on trucks. The solid rubber was tough, but provided a rough ride (Dyke, *Automobile and Gasoline Engine*, pp. 560-561; Richard Sutton, *Car* [New York: Alfred A. Knopf, 1990], p. 60).

5. *San Antonio Express*, November 1, 1936.

6. Gertrude Fischer Interview, March 6, 1989; Mengden Interview.

7. The Devil's Backbone drive followed the same route as present-day Ranch Road 32.

8. *San Antonio Express*, September 4, 1938. The author will refer to the stop as Fischer's Store, the way it appeared on the scenic county map, unless specifically used otherwise in quotes or titles. In 1989 the author viewed a faded, tattered copy of Joe Sanders's 1933 county road map, still tacked on the back wall of the store. The old Fischer Store remained open until November 27, 1995 (*New Braunfels Herald-Zeitung*, December 20, 1995).

9. Sanders Interview, August 9, 1990; Fischer Interview.

10. *San Antonio Express*, September 4, 1938.

11. Com. Ct., Vol. M, p. 632; *San Antonio Express*, September 4, 1938.

12. In the 1990s North Cranes Mill Road began off Ranch Road 32 at Fischer Store.

13. Kolodzie Interview; Roger Nuhn, ed., *The New Braunfels Sesquicentennial Minutes* (New Braunfels: Sophienburg Museum and Archives, 1995), pp. 42, 117. The area of Startz Hill had been known as Hillview in earlier years. In 1950 the roadside stop became a Comal County park and is now located off Farm Road 3159.

14. *San Antonio Express*, November 1, 1936.

15. Jasinski Interview, August 21, 1991; Sanders Interviews, August 25, 1988, January 1989; Kaderli Interview; Mengden Interview.

16. *San Antonio Express*, November 1, 1936.

17. Roth Interview; *Road Map of Comal County*, 1936.

18. *San Antonio Express*, November 1, 1936.

19. Ibid., July 23, 1933.

TEN: NATURE AT WORK: SECRET PLACES, CAVES, AND CREATURES

1. *San Antonio Express*, November 1, 1936.
2. Ibid. The terms "cave" and "caverns" involve different characteristics in size to cavers. Some also define caverns as "living," containing active, growing formations, and caves as usually dry or "dead" chambers. The term "cave" was often freely used in advertisements and write-ups in the 1930s, and the author has chosen to include this term as a general synonym regarding "living" caverns.
3. Sanders Interview, August 14, 1988.
4. *San Antonio Express*, July 23, 1933. For a more extensive history of Jacob's Well see: Laurie E. Jasinski, "Spring Power," *Texas Parks and Wildlife*, Vol. 48, No. 8 (August 1990), pp. 20–27.
5. Ibid.
6. Sanders Interview, August 25, 1988.
7. *San Antonio Express*, July 10, 1932.
8. "Fairy Cavern," tourist brochure, ca. early 1930s (author's possession).
9. Cascade Caverns officially opened in 1932 (Jerry and Dorothy Sinise, *Texas Show Caves* [Austin: Eakin Press, 1983], p. 47; *Texas State Travel Guide* [Austin: Texas Department of Transportation, 1995], p. 64; phone call to Cascade Caverns, 1997).
10. The cave, also known as Bracken Bat Cave, is home to as many as twenty million bats. In the 1990s gardening business interests still harvested guano from Bracken Bat Cave, and nurseries such as the Garden-Ville chain sold the product. The bat cave and four surrounding acres were owned by Bat Conservation International, a group based in Austin (*San Antonio Express-News*, February 16, 1997).
11. Also see Mark McKenna, "Some Good Ole Rebel Ingenuity: The Guano Oven of New Braunfels, Texas," *Blue & Gray* (June 1989), p. 62. The guano oven of New Braunfels is located in Landa Park, and a granite marker honors the site. The Confederacy's Nitro and Mining Department produced saltpeter from the heated guano.
12. Sanders Interview, April 6, 1988.
13. Otto Locke, Jr., Interview, July 19, 1990; Oscar Haas, "J. J. Locke Saw Need for Nursery in 1856," *Southern Florist and Nurseryman*, Vol. 78 (July 16, 1965), n.p.; *The Texas Nurseryman*, "Otto Locke Nursery Since 1856," Vol. 10, No. 4 (April 1979), pp. 4–5; Sanders Interview, April 6, 1988.
14. Maria Liebscher and Franziska Dittlinger Liebscher Interview, February 19, 1990. Maria, Carl, and Loretta Liebscher are the children of Alfred and Franziska Liebscher.
15. Sanders Interview, April 6, 1988. The story was covered in the *San Antonio Express* (July 25, 1937), *New Braunfels Herald* (July 18, 1937, July 23, 1937), and *Victoria Advocate* (July 18, 1937). Crimmins and Gloyd corresponded extensively

with Mrs. Sanders. Joe Sanders conferred with Willis Wollens of the reptile garden at Brackenridge Park in San Antonio and also with Mrs. Ellen Quillin, director of the Witte Museum. The snake was sent (alive) for study at the Chicago Academy of Sciences and eventually went on display at the Witte Museum in San Antonio. In 1988 the author called the museum regarding the preserved coral snake, but the specimen could not be found.

Dr. Howard K. Gloyd, director of the Chicago Academy of Sciences, published a journal article: "A Case of Poisoning From the Bite of a Black Coral Snake," *Herpetologica*, Vol. 1 (1938), pp. 121-124. This article offers the most accurate account of the incident. Some newspapers erroneously reported that Mrs. Sanders stuck her hand in a snake box and was bitten.

16. Sanders Interview, August 14, 1988; Clara Wuest Heidemann Interviews, May 17, 1988, June 8, 1988; Reginald Wuest Interview, March 17, 1988. Joe was correct about a cave site at the natural bridge. In 1960, four college students from St. Mary's University in San Antonio explored the cave opening beyond the first chamber and discovered a huge underground cavern. After excavation and development Natural Bridge Caverns opened in 1964. Advertised as Texas' largest caverns, by the 1980s cavers had mapped some two miles.

ELEVEN: SCENIC SOUVENIRS

1. Sanders Interview, April 1996; Naylor, *Naylor's Epic-Century Magazine*, pp. 5-9, 64; "Night Scenes of Texas Centennial Exposition, Dallas, 1836-1936," postcard booklet, Dallas: Dallas Post Card Co., Made in C. T. Art-Colortone by Curt Teich & Co., Chicago, 1936 (author's collection). The Sanders family left for the Texas Centennial on October 10, 1936. They were accompanied by Mrs. Alfred (Franziska) Liebscher, her children Carl and Loretta, Mrs. Dittlinger, and her son Bruno. The Texas Centennial was so popular that organizers brought it back in spring 1937, and it was held over for several months.

2. *New Braunfels Herald*, October 23, 1936, February 19, 1937; Schumann, "The Great Depression," pp. 109-111. The original twenty road signs were to measure four feet by eight feet at an expense of $500.

3. Sanders Interviews, August 17, 1990, August 23, 1990; Rudy Seidel Interview, April 9, 1990.

4. Seidel Interview. The Sophienburg Archives houses the negatives from Seidel's studio. The collection covers years from about 1929 to 1970.

5. Rudy Seidel believed that the postcard camera used 122m.m. film.

6. Sanders Interviews, August 17, 1990, April 1996. Several scenes were actually from photos taken some years earlier. One Guadalupe River Drive photo dated to approximately 1932. The scene shows the distant figures of Laurie and her daughter, about seven years old at the time. The postcards were numbered. The highest number in the author's possession is 43. Hip Mengden remembered a series total of 50 cards. The postcards had "Seidel Studio, New Braunfels, Tex." stamped on the back.

7. Legion Minutes, August 3, 1939. The author does not know what services Philip Rawson provided. According to Rudy Seidel, the postcards were sold for a wholesale price of a quarter of a cent each, and stores in turn charged a slightly higher retail price.

8. Com. Ct., Vol. P, pp. 211, 220, 349.

9. Legion minutes, April 14, 1941; *The American Legion Scenic Road Map of Comal County Texas,* Comal Post 179, New Braunfels, Compiled by Joe J. Sanders, Drawn by Roye V. Swartwout, 1941, Sophienburg Archives, New Braunfels, Texas. Cave Without A Name opened in 1939. The owner held a contest to name the cavern. A young boy's suggestion, Cave Without a Name, was the winner. He had commented that the place was too pretty to have a name (Morris Hastings, "Cave Without A Name," *Texas Parade,* Vol. 4, No. 12 [May 1940], pp. 12-13, 26).

10. *Road Map of Comal County* ... 1941; Kearney Interview. Highway 281 was also known as the American Legion Memorial Highway. The golf course adjacent to Landa Park opened in summer 1940.

11. The three summer resorts were Ulbricht's Resort, Camp Warnecke, and Camp Giesecke. The 1940 United States census recorded a population of 6,976 for New Braunfels. In 1943 the town had an estimated population of 7,500 (*Texas Almanac,* 1943-1944, p. 80). In 1941 New Braunfels included 210 listed businesses (*Texas Almanac,* 1941-1942, p. 126).

12. Joe Sanders Legion Scrapbook, undated newspaper clipping.

13. *New Braunfels Herald,* August 22, 1941. The WPA was one of Franklin Roosevelt's "New Deal" programs and part of legislation in 1935 to provide work for the unemployed. Jobs included not only employment for laborers to build roads and bridges, but also positions for writers, artists, and musicians to work on cultural pursuits. The organization originally was called the Works Progress Administration, but the middle name later became "Projects" (*Family Encyclopedia of American History,* "New Deal," p. 787, "Works Progress Administration," p. 1261).

TWELVE: IS THIS TRIP NECESSARY?

1. Lewis and Musciano, *Automobiles,* pp. 430-432; *Motorized America In War and Peace* (Washington, D. C.: Automotive Safety Foundation, 1945), pages not numbered. Subtitle reads: "A picture story of how the motor vehicle and the road helped to bring victory over Germany and Japan, and a glimpse of today's and tomorrow's highway transportation."

2. Ibid.; Sanders Interviews, August 14, 1990, May 15, 1996, May 17, 1996. Initially, gasoline rationing was enforced in seventeen eastern states on May 15, 1942. Food and building materials were also rationed, and many people used food rationing stamps for items like sugar or meat. Local farmers shared much of their excess meat or meat stamps with needy consumers.

3. *Motorized America,* n.p.; Sanders Interviews, May 15, 1996, May 17, 1996. According to *Motorized America,* at the end of 1945 cars that were ten years old or more made up thirty-four percent of all cars on the road. Cars aged seven to

ten years made up thirty-two percent; cars aged three to seven years made up thirty-one percent. Only three percent of the cars were three years old or less.

4. Lewis and Musciano, *Automobiles*, p. 433; *Motorized America*, n.p.

5. Joe marked each sign according to its numbered intersection (intersection 14, for example) and then with a letter (A, B, C, etc.) depending on how many signs were posted at that crossroads. (One intersection had five signs.) In this manner, the schedule enabled one to locate easily a specific sign at a specific crossroads.

6. Rice Interview; Kearney Interview, March 2, 1989; *San Antonio Express and News*, February 16, 1963. Years later after Clarence Rice had left the county judge's office, in the process of cleaning and renovation, many items filed were put in permanent storage and others were destroyed. During this time, Larry Kearney of the Comal County Road Department discovered the Sanders road schedule and saved it. For some years the compilation was housed in a filing cabinet in the headquarters of the Comal County Road Department until one day the author met Mr. Kearney, who, while relaying road information, happened to bring forth the Sanders schedule. A second copy is now filed at the Comal County Road Department headquarters.

7. Sanders issued another revised map with Swartwout in 1948. In the late 1940s Joe still actively participated in highway beautification and wrote an article for *Texas Legion News* urging the Legion to do its part (J. J. Sanders, "Hiway Beautification Aid Asked of Legion," *Texas Legion News*, Vol. 19, No. 6 [March 25, 1947], p. 6).

8. "New Braunfels, 'The Beauty Spot of Texas,' Centennial Celebration," Official Souvenir Program, May 10-12, 1946, Center for American History, University of Texas at Austin; *Road Map of Comal County* ... 1946. Actually, the town turned 100 in 1945, but the celebration was postponed a year because of the war and due to a devastating storm that hit Landa Park.

9. "Highway Highlights" bulletins, 1-30-46, p. 4; 8-27-46, p. 1; 11-26-46, p. 1; 3-29-47, p. 3; 4-24-47, p. 1; 6-19-47, p. 1, Center for American History, University of Texas at Austin; "State Department of Highways and Public Transportation, Introduction," ca. 1986; Com. Ct., Vol. Q, p. 236. The total of 26,864 miles does not include the 7,205 miles of farm-to-market roads designated by the Texas State Highway Commission prior to September 1945 since the farm road program was still in its infancy. The amended total of state highways and farm-to-market roads was 34,069 miles. Actually, as early as May 1943 the Forty-eighth Texas Legislature had given the state highway commission the authority to name any county road in Texas as a farm-to-market road for construction and maintenance purposes. Acts of the Fiftieth Texas Legislature resulted in a cooperative road plan in which a county or road district furnished rights-of-way and seventy-five percent initial construction costs while the highway department supplied the engineering, twenty-five percent construction costs, and then assumed responsibility for perpetual maintenance.

   In April 1947 the gasoline tax was 5.5 cents per gallon—4 cents state and 1.5 cents federal tax. In 1996 the gasoline tax in Texas was 38.4 cents per gallon—20 cents state and local tax and 18.4 cents federal tax (source: Chevron gasoline station in New Braunfels).

10. "Highway Highlights," 11-15-47, p. 1.

11. Com. Ct., Vol. Q, pp. 226, 236-238, Vol. R, pp. 13, 36.

12. *Texas Highways*, pp. 132-135; Kolodzie Interview. For modern county roads, crews allow for a maximum of a ten percent up or down grade. Hence, on rugged, knobby terrain deep cuts in the hillside are often seen along the roadsides.

13. Kearney Interview; *Austin American*, March 11, 1958; Kolodzie Interview; Com. Ct., Vol. Q, p. 33, Vol. R, pp. 22, 36, 101-102, 173-174, 211, 560, 618-619, Vol. S, pp. 67, 190-191. Many water lanes (explained in Chapter One) in the county were also abandoned. The Rebecca Creek School, organized in 1882, was consolidated with other rural schools to form the Sherwood School District in 1944 (Nuhn, *New Braunfels*, pp. 130, 136). In Texas Interstate Highway 35, which crosses the Red River north of Gainesville and runs south to Laredo, was under construction in 1958. In New Braunfels the freeway replaced U.S. 81, which became a major city avenue. Engineers constructing interstate highways generally allow roughly a three percent up or down grade.

14. *Austin American*, March 11, 1958; "Plan For Highway Development," Booklet, New Braunfels, Texas, 1958, pp. 11, 16, Center for American History, University of Texas at Austin; "Texas and the Interstate Highway System," Booklet, Austin: Texas Highway Department, 1957, pp. 3-5; *Texas Highways*, 1967, pp. 32-33. From January 1, 1946, to approximately March 1958, 241 miles of state highways were abandoned. The federal government conceived of the interstate highway system in 1944 for national defense purposes. In 1956 the Federal Aid Highway Act provided funding for the National System of Interstate and Defense Highways to connect major cities across the United States.

15. County surveyor William Kolodzie conducted an extensive survey of River Road in the mid-1970s. Surveyors determined the center line, and the county claimed a forty-foot right-of-way, even though many sections had only a thirty-foot right-of-way. In 1991 Kolodzie recalled that one of the Sanders road signs was still posted on Krueger Lane some twenty or so years earlier.

16. Sanders Interview, January 1989; Harvey Ruppel Interview, July 15, 1991. H. Dittlinger died on September 29, 1946. His wife had died December 31, 1943. Laurie Jo Sanders married Richard Jasinski in 1954. They had two children, Lawrence in 1957 and Laurie Eileen in 1964. Harvey Ruppel, a sign painter and owner of Ruppel Signs, recalled that in the later 1940s and 1950s Sanders and the Legion Post contracted out some sign work to his father Bill "Rusty" Ruppel. Bill charged perhaps $1.00 for each newly painted sign.

17. Sanders Interview, January 1989; *Road Map of Comal County* ... 1955. Joe examined topographical maps and on his own tried to determine a precise location for a dam. The 1936 Comal County map contained a shoreline of a proposed flood-control lake, but at that time a dam site was placed miles downstream, near Hueco Springs, and the reservoir would have covered the River Road area. By the 1950s the proposed site had been moved about sixteen miles upriver.

18. Sanders Interview, August 9, 1990; Kearney Interview. From where North Cranes Mill Road meets Fischer Store Road at Ranch Road 32, it goes approximately five

miles south and then runs into the lake. South Cranes Mill Road picks up off Farm Road 3159 and proceeds about four miles to Highway 46. Larry Kearney commented on the remarkable conditions of Hancock Road, even after over two decades of submersion. During a dry spell, parts of the old road emerged and he was able to drive on them.

19. "Plan For Highway Development," p. 15; *Comal County Road Maps*, 1951 and 1955; Kearney Interview; General Highway Map, Comal County, Texas, State Department of Highways and Public Transportation, Austin, Texas, 1988; Sanders Interview, January 1989. According to Laurie Sanders, to a small degree the purchase of land tracts began as early as the 1930s. During the Depression some wealthy investors bought land from farmers and ranchers who needed the money. Land was very inexpensive—as cheap as $10 an acre. Buying real estate accelerated after World War II as more people moved to the area, and the construction of Canyon Dam and Reservoir greatly encouraged the market. According to Larry Kearney, by the 1980s and 1990s subdivisions made up the majority of new roads under construction in the county. The old class system of roads (first, second, and third class) had been replaced with new categories: arterial roads and subdivisions.

20. Larry Jasinski Interview, June 4, 1996. Larry Jasinski recalled accompanying his grandfather to a place downtown on South Seguin Street, probably the offices of the *New Braunfels Zeitung-Chronicle*, where they picked up newly printed copies of the 1960 Comal County map.

21. Rice Interview; Kearney Interview; Eddie Gumbert, conversation, February 11, 1988; Bodo Dietert, Telephone Interviews, March 14, 1988, February 18, 1989; David Jonas, conversations, February 21, 23, 24, 1989. Construction of Canyon Dam and Lake began in April 1958; impoundment began in June 1964 ("Canyon Dam and Lake," brochure, Corps of Engineers, U.S. Army, Southwestern Division, Fort Worth District). In 1991 Comal County had approximately 685 miles in the county road system, and an average of seventy miles were repaired each year (*New Braunfels Herald-Zeitung*, June 20, 1991).

    The author is fortunate to have two of the road signs. Their discovery tells a remarkable story. After over a year of searching, the author had contacted retired and current road department employees, and it was becoming apparent that the signs had long been taken or destroyed. Finally, road employee David Jonas found two signs at the old road warehouse in Bulverde. Previously the county had operated four equipment storage facilities for the four road precincts in the county. The Bulverde site was the last of the road warehouses still open. The signs were nailed to the wall and had been covered by other stored items. Both signs were originally posted in the Bulverde area near the Comal County-Bexar County line. Mr. Jonas contacted the author and took the signs to the road headquarters.

22. *Texas Almanac ...* 1964–1965, p. 467; *Texas Almanac ...* 1978–1979, p. 548; *State Highway Department of Texas, Twenty-Fifth Biennial Report*, September 1, 1964 to August 31, 1966, Austin, Texas, p. 13, Albert B. Alkek Library, Southwest Texas State University, San Marcos, Texas; *Road Map of Scenic Comal County Texas And Adjacent Areas*, Originally compiled by J. J. Sanders for Comal Post 179, American Legion, Drawn by Phil Rawson, Printed and distributed by Comal County

Chamber of Commerce and New Braunfels Board of City Development, New Braunfels, Texas, 1960, Sophienburg Archives, New Braunfels, Texas; *Texas Almanac* ... 1961-1962, p. 213.

In 2000 there were over 296,000 miles of roads in Texas. As of March 2000, the Texas Department of Transportation maintained 79,102 miles. This total included 3,233 miles of interstate highways, 6,402 miles of interstate frontage roads, 16,419 miles of state highways, 12,110 miles of U.S. highways, and 40,938 miles of farm-to-market roads. The total mileage maintained by the state does not include county and city-controlled roads. In 1999 over 17 million vehicles were registered in Texas ("Pocket Facts," Austin: Texas Department of Transportation, March 2000). Among the revisions suggested by Joe Sanders was a change in the title from *Scenic Road Map* to *Road Map of Scenic Comal County*.

23. Laurie Sanders Jasinski Interview, August 21, 1991. The Narrows closed in the 1960s. In the 1980s Jacob's Well was on private property in the Woodcreek development and closed to the public. Over the years a number of scuba divers lost their lives there while trying to trace the origin of the well spring. Edge Falls closed in the 1970s after a drowning. Mrs. Sanders and the Jasinski family visited Edge Falls around 1967 but were shocked at the amount of litter strewn about the overcrowded park.

24. By the 1970s the roadside picnic was a thing of the past along River Road. Since the impoundment of Canyon Lake, part of the river valley underwent a gradual transformation to a more commercial-oriented area. Petitions to improve River Road came before the commissioners court in 1965 (Vol. U, p. 447), and on May 24, 1966, the court ordered that all of River Road be accepted for county maintenance (Vol. U, p. 598). In October 1967 a motion carried to urge Congress to consider the Guadalupe River for future legislation in its Wild and Scenic Rivers bill (Vol. V, p. 265). In August 1977, the commissioners court recorded a resolution of opposition to include the Guadalupe River for study and/or inclusion in the Wild and Scenic Rivers System (Vol. Y, p. 447). By this time various parking petitions were being introduced as traffic became heavier (Vol. Y, pp. 483, 524-525). The law was eventually passed. In 1980 signs costing $526 were erected along River Road (Vol. Z, p. 533). In the 1980s and 1990s much of River Road had commercial rafting and camping establishments and catered to the summer tourist business. River Road also had the highest number of auto accidents and claims (ca. 1990) against the county—close to ten times that of any other road (Kearney Interview). Many accidents were alcohol-related.

25. *San Antonio Express and News*, February 16, 1963.

26. Letter from A. W. Grant, Austin, Texas, to the Sanders family, New Braunfels, Texas, December 19, 1963 (author's possession). Joe Sanders died on June 20, 1964.

BIBLIOGRAPHY

NEWSPAPERS AND PERIODICALS

Austin American. April 3, 1924; March 11, 1958.

Blanco County News. April 23, 1936.

Dallas Morning News. April 10, 1904; May 24, 1936.

Houston Chronicle. April 19, 1936.

New Braunfels Herald. May 8, 1914; October 23, 1914; April 10, 1925; July 24, 1925; October 2, 1925; November 5, 1926; March 20, 1931; March 27, 1931; April 3, 1931; May 8, 1931; June 23, 1933; July 14, 1933; July 28, 1933; August 25, 1933; December 1, 1933; December 8, 1933; December 29, 1933; May 18, 1934; May 25, 1934; July 13, 1934; February 1, 1935; March 22, 1935; April 19, 1935; August 16, 1935; April 10, 1936; October 23, 1936; February 19, 1937; July 18, 1937; July 23, 1937; October 15, 1937; August 22, 1941; December 31, 1964.

New Braunfels Herald-Zeitung. June 20, 1991, December 20, 1995; March 28, 1996.

New Braunfels Zeitung-Chronicle. April 12, 1964.

San Antonio Daily Express. May 1, 1904.

San Antonio Express. July 25, 1912; October 8, 1913; October 4, 1925; October 29, 1929; April 26, 1931; July 10, 1932; July 23, 1933; November 1, 1936; July 25, 1937; August 21, 1938; September 4, 1938.

San Antonio Express and News. February 16, 1963; September 26, 1965.

San Antonio Express-News. February 16, 1997.

Sanders, J. J. "Hiway Beautification Aid Asked of Legion." *Texas Legion News*, Vol. 19, No. 6 (March 25, 1947), p. 6.

San Marcos Record. January 9, 1920.

Victoria Advocate. July 18, 1937.

Victoria Daily Advocate. 1922.

INTERVIEWS CONDUCTED BY LAURIE E. JASINSKI

Barnett, William. March 15, 1988, Comal County, Texas, Seven Eleven Ranches.

Fischer, Gertrude. March 6, 1989, Fischer Store, Texas.

Heidemann, Clara Wuest. May 17, 1988; June 8, 1988, Natural Bridge Caverns, Comal County, Texas.

Henk, Erwin. February 21, 1991, New Braunfels, Texas.

Jasinski, Larry. June 4, 1996, New Braunfels, Texas.

Jasinski, Laurie Sanders. August 14, 1988; August 21, 1991, New Braunfels, Texas.

Jasinski, Laurie Sanders and Laurie P. Sanders. August 9, 1990; August 21, 1991, New Braunfels, Texas.

Kaderli, Milton. March 11, 1991, New Braunfels, Texas.

Kearney, Larry. December 5, 1991, Comal County Road Department, New Braunfels, Texas.

Kerbow, Dorothy Wimberley. February 14, 1989, Wimberley, Texas.

Kolodzie, William. September 16, 1991, New Braunfels, Texas.

Liebscher, Franziska Dittlinger and Maria Liebscher. February 19, 1990, New Braunfels, Texas.

Locke, Otto, Jr. July 19, 1990, New Braunfels, Texas.

Mengden, Hip. August 9, 1991, New Braunfels, Texas.

Nowotny, E. P. "Pete." June 24, 1991, New Braunfels, Texas.

Riley, J. C. February 17, 1989, Hunter, Texas.

Ruppel, Harvey. July 15, 1991, New Braunfels, Texas.

Sanders, Laurie Pfau. April 6, 1988; July 21, 1988; August 14, 1988; August 25, 1988; Winter 1989; January 1989; Summer 1990; August 9, 1990; August 13, 1990; August 14, 1990; August 17, 1990; August 23, 1990; February 21, 1991; Spring 1991; June 30, 1991; October 1991; October 31, 1991; August 27–28, 1992; Fall 1992; January 1993; January 23, 1993; Summer 1993; July 1993; Winter 1994; Spring 1994; March 6, 1995; April 1996; May 15, 1996; May 17, 1996; October 31, 1996; New Braunfels, Texas.

Sanders, Laurie P. and Ralph Pfau. January 8, 1989, New Braunfels, Texas.

Seidel, Rudy. April 9, 1990, New Braunfels, Texas.

Stahl, Wally. February 28, 1991, New Braunfels, Texas.

Wuest, Reginald. March 17, 1988, Comal County, Texas, Natural Bridge Caverns.

Wyatt, Tula Townsend. February 18, 1989, San Marcos, Texas.

TELEPHONE INTERVIEWS WITH THE AUTHOR

Dietert, Bodo. March 14, 1988; February 18, 1989.

Owens, Jim. August 4, 1991.

Rice, Clarence. February 21, 1989.

PERSONAL CONVERSATIONS WITH THE AUTHOR

Boyd, Charlotte. Comal County Clerk's Office, June 1991, New Braunfels, Texas.

Gumbert, Eddie. February 11, 1988, Wimberley, Texas.

Jonas, David. February 21, 23, 24, 1989, New Braunfels, Texas.

Kearney, Larry. March 2, 1989, New Braunfels, Texas.

BOOKS

Brokaw, H. Clifford and Charles A. Starr. *Putnam's Automobile Handbook*. New York: G. P. Putnam's Sons, The Knickerbocker Press, 1918.

Burness, Tad. *Cars of the Early Twenties*. Philadelphia: Chilton Book Company, 1968.

Clymer, Floyd. *Henry's Wonderful Model T, 1908-1927*. New York: Bonanza Books, 1955.

Day, Richard. *The Practical Handbook of Concrete and Masonry*. New York: Fawcett Publications, Inc., 1969.

Dyke, A. L. *Dyke's Automobile and Gasoline Engine Encyclopedia*. Twelfth Edition. St. Louis: A. L. Dyke, Publisher, 1920.

Family Encyclopedia of American History. Pleasantville, New York: The Reader's Digest Association, Inc., 1975.

Fuchs, John R. *A Husband's Tribute To His Wife*. San Antonio: The Naylor Company, 1938.

General Laws of the State of Texas Passed at the Second Session of the Fourteenth Legislature. Houston: A. C. Gray, State Printer, 1875.

Georgano, G. N., ed. *The Complete Encyclopedia of Motorcars, 1885 to the Present*. New York: E. P. Dutton and Company, Inc., 1968, 1973.

Haas, Oscar. *History of New Braunfels and Comal County, Texas, 1844-1946*. Austin: Hart Graphics, Inc., 1968.

Howell, Mark. *Racing Stutz*. New York: Ballantine Books, 1972.

Landa, Harry. *As I Remember. . . .* San Antonio: Carleton Printing Company, 1945.

Lewis, Albert L. and Walter A. Musciano. *Automobiles of the World*. New York: Simon and Schuster, 1977.

Motorized America In War and Peace. Washington, D.C.: Automotive Safety Foundation, 1945.

Nuhn, Roger, ed. *The New Braunfels Sesquicentennial Minutes*. New Braunfels: Sophienburg Museum and Archives, 1995.

Roberts, Peter. *A Pictorial History of the Automobile*. New York: Grosset and Dunlap, 1977.

Sinise, Jerry and Dorothy Sinise. *Texas Show Caves.* Austin: Eakin Press, 1983.

Sutton, Richard. *Car.* Eyewitness Books Series. New York: Alfred A. Knopf, 1990.

Texas Highway Department. *Eleventh Biennial Report of the State Highway Department.* September 1, 1936, to August 31, 1938, Austin: Texas Highway Department, 1938.

_____. *State Highway Department of Texas, Seventh Biennial Report.* September 1, 1928, to August 31, 1930, Austin: Texas Highway Department, 1930.

_____. *State Highway Department of Texas, Twenty-Fifth Biennial Report.* September 1, 1964, to August 31, 1966, Austin: Texas Highway Department, 1966.

The Texas Almanac and State Industrial Guide. Dallas: A. H. Belo and Company, 1925; 1936; 1941-1942; 1943-1944; 1961-1962; 1964-1965; 1978-79.

Texas State Travel Guide. Austin: Texas Department of Transportation, 1995.

Tyler, Ron, ed. *The New Handbook of Texas.* 6 vols. Austin: Texas State Historical Association, 1996.

Verrill, A. Hyatt. *How To Operate a Motor Car.* Philadelphia: David McKay, Publisher, 1918.

Webb, T. H., compiler, Assistant State Highway Engineer. *Texas State Highway Commission: Maintenance Manual Inaugurating the Patrol System of Maintenance.* Austin: Texas State Highway Commission, Published for the Information of County Commissioners, County Road Superintendents and Patrolmen, 1922.

"REFLECTIONS," KGNB/KNBT RADIO PROGRAMS

THE FOLLOWING TAPES AND TRANSCRIPTS CAN BE FOUND AT THE SOPHIENBURG ARCHIVES, NEW BRAUNFELS, TEXAS.

Hillje, Jarvis Gregory. "Reflections" No. 124. Interview conducted by Herb Skoog, aired April 22, 1979.

Hoffmann, Walton. "Reflections" No. 467. Interview conducted by Herb Skoog, aired March 22, 1987.

Jarisch, Egon (and wife Ella Jarisch). "Reflections" No. 177. Interview conducted by Herb Skoog, aired April 27, 1980.

Liebscher, Alfred. "Reflections" No. 30. Interview conducted by Herb Skoog (with Fred Oheim), aired August 7, 1977.

Oheim, Fred. "Reflections" No. 1. Interview conducted by Herb Skoog, aired December 12, 1976.

Roth, Melba Meckel. "Reflections" No. 335. Interview conducted by Herb Skoog, aired November 13, 1983.

Ruppel, John B. "Reflections" No. 26. Interview conducted by Herb Skoog (with Fred Oheim), aired June 5, 1977.

MAGAZINES AND JOURNALS

The American Legion Monthly. Vol. 14, No. 3 (March 1933), p. 32.

Automobile Trade Journal. New York Silver Jubilee Show Number, Chilton Company, Philadelphia, Vol. 24, No. 7 (January 1, 1925), pp. 111, 200.

Gloyd, Howard K. "A Case of Poisoning From the Bite of a Black Coral Snake." *Herpetologica*, Vol. 1 (1938), pp. 121-124.

Haas, Oscar. "J. J. Locke Saw Need for Nursery in 1856." *Southern Florist and Nurseryman*, Vol. 78 (July 16, 1965), n.p.

Hastings, Morris. "Cave Without A Name." *Texas Parade*, Vol. 4, No. 12 (May 1940), pp. 12-13, 26.

The Inland Merchant. "Younger Every Year." March, 1924, pp. 27-28, in "Somers V. Pfeuffer" Vertical File, Center for American History, University of Texas at Austin.

Jasinski, Laurie E. "Memories of Early Sarita." *The Journal of South Texas*, South Texas Historical Association, Vol. 6, No. 1 (1993), pp. 24-53.

_____. "Spring Power." *Texas Parks and Wildlife*, Vol. 48, No. 8 (August 1990), pp. 20-27.

Kellman, Jerold L., ed. *Cars of the 30s*. Skokie, Illinois: *Consumer Guide*, Classic Car Bi-Monthly, Vol. 269 (May 1980).

Lively, Frank, ed. *Texas Highways*. Fiftieth Anniversary Edition, Austin: Texas Highway Department, Travel and Information Division, Vol. 14, No. 9 (September 1967).

Maddox, Frank. "Know the Roads You Ride On." *Texas Parade*, Vol. 1, No. 2 (July 1936), p. 7.

McKenna, Mark. "Some Good Ole Rebel Ingenuity: The Guano Oven of New Braunfels, Texas." *Blue & Gray*, June 1989, p. 62.

Naylor, Joe O. *Naylor's Epic-Century Magazine: The Illustrated Historical Quarterly of Texas*, Vol. 3, No. 1 (April 1936), pp. 5-9, 64.

Simons, Charles E. "Traveler's Oasis: The Story of Texas' Roadside Parks." *Texas Parade*, Vol. 1, No. 4 (September 1936), p. 3.

The Texas Nurseryman. "Otto Locke Nursery Since 1856." Vol. 10, No. 4 (April 1979), pp. 4-5.

MAPS

(Map citations for 1920-21, 1927, 1933, and 1936 can be found at the Center for American History, University of Texas at Austin. Unless otherwise noted, all others are housed at the Sophienburg Archives, New Braunfels, Texas.)

Sanders, J. J., compiler. *Road Map of Comal County Texas*. Comal Legion Post 179, Compiled by J. J. Sanders, Drawn by C. H. Bernstein, 1933.

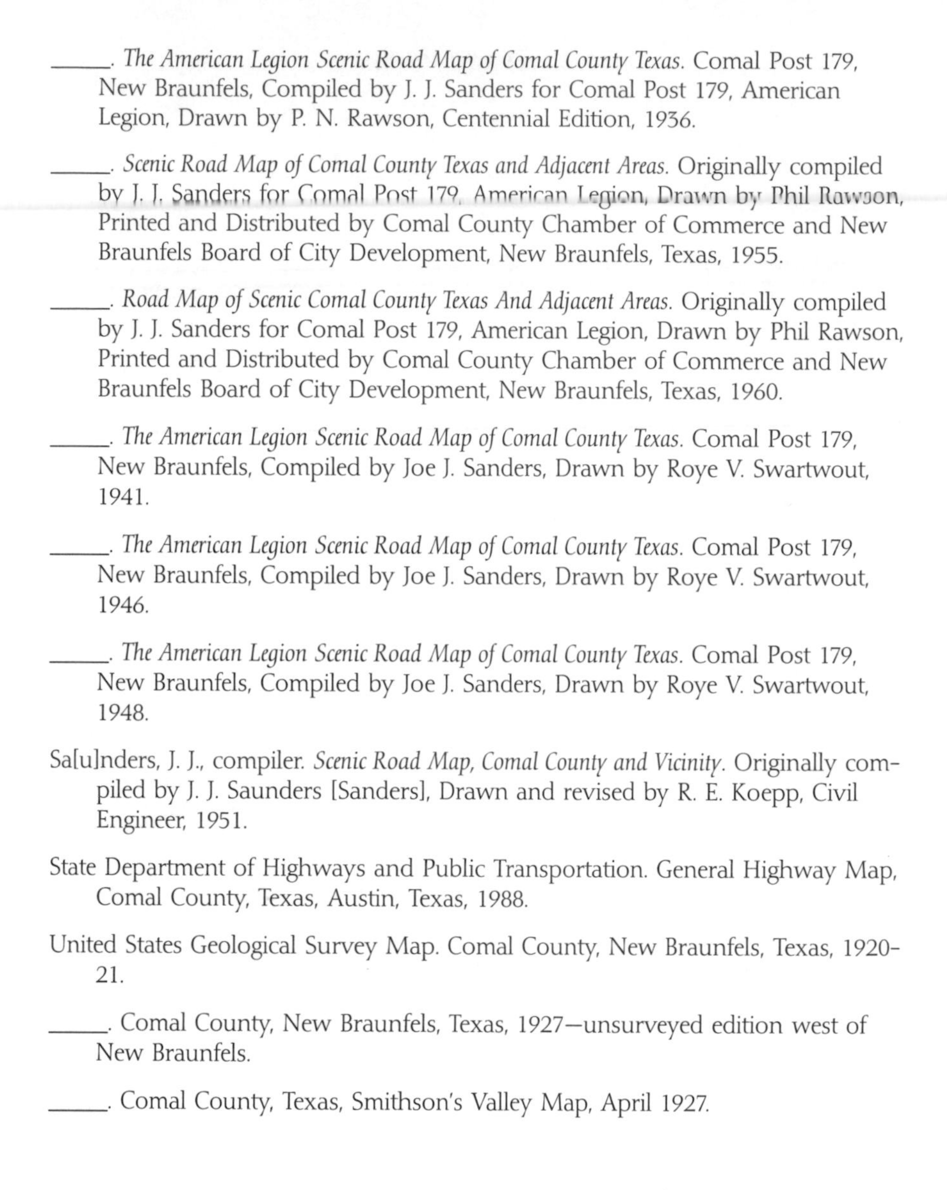

______. *The American Legion Scenic Road Map of Comal County Texas.* Comal Post 179, New Braunfels, Compiled by J. J. Sanders for Comal Post 179, American Legion, Drawn by P. N. Rawson, Centennial Edition, 1936.

______. *Scenic Road Map of Comal County Texas and Adjacent Areas.* Originally compiled by J. J. Sanders for Comal Post 179, American Legion, Drawn by Phil Rawson, Printed and Distributed by Comal County Chamber of Commerce and New Braunfels Board of City Development, New Braunfels, Texas, 1955.

______. *Road Map of Scenic Comal County Texas And Adjacent Areas.* Originally compiled by J. J. Sanders for Comal Post 179, American Legion, Drawn by Phil Rawson, Printed and Distributed by Comal County Chamber of Commerce and New Braunfels Board of City Development, New Braunfels, Texas, 1960.

______. *The American Legion Scenic Road Map of Comal County Texas.* Comal Post 179, New Braunfels, Compiled by Joe J. Sanders, Drawn by Roye V. Swartwout, 1941.

______. *The American Legion Scenic Road Map of Comal County Texas.* Comal Post 179, New Braunfels, Compiled by Joe J. Sanders, Drawn by Roye V. Swartwout, 1946.

______. *The American Legion Scenic Road Map of Comal County Texas.* Comal Post 179, New Braunfels, Compiled by Joe J. Sanders, Drawn by Roye V. Swartwout, 1948.

Sa[u]nders, J. J., compiler. *Scenic Road Map, Comal County and Vicinity.* Originally compiled by J. J. Saunders [Sanders], Drawn and revised by R. E. Koepp, Civil Engineer, 1951.

State Department of Highways and Public Transportation. General Highway Map, Comal County, Texas, Austin, Texas, 1988.

United States Geological Survey Map. Comal County, New Braunfels, Texas, 1920–21.

______. Comal County, New Braunfels, Texas, 1927—unsurveyed edition west of New Braunfels.

______. Comal County, Texas, Smithson's Valley Map, April 1927.

LETTERS

Allred, James V., Governor of Texas, Austin, Texas, to J. J. Sanders, New Braunfels, Texas, February 29, 1936. Letter in author's possession.

Buske, Mayme, New Braunfels, Texas, to Joe Sanders, New Braunfels, Texas, undated—approximately April 1932. Letter in Joe Sanders Legion Scrapbook in author's possession.

Grant, A. W., Austin, Texas, to the Sanders family, New Braunfels, December 19, 1963. Letter in author's possession.

Jones, W. G., Secretary-Manager, Motor League of South Texas, Houston, Texas, to J. J. Sanders, New Braunfels, Texas, July 22, 1933. Letter in Joe Sanders Legion Scrapbook in author's possession.

SCRAPBOOKS, DIARIES, BROCHURES, COMPILATIONS, RESEARCH FILES

"Canyon Dam and Lake." Brochure, Corps of Engineers, U.S. Army, Southwestern Division, Fort Worth District, n.d, in author's possession.

"Fairy Cavern." Tourist Brochure, ca. early 1930s in author's possession.

"Highway Highlights." Bulletins, Austin: Texas Good Roads Association, 1-30-46; 8-27-46; 3-29-47; 4-24-47; 6-19-47; 11-15-47 Center for American History, University of Texas at Austin.

"History of Texas Roads and the Texas Highway Department." Austin: Texas Highway Department, Traffic Services Division, ca. 1948, Center for American History, University of Texas at Austin.

Hoffmann, Walton F., compiler. "Birth of the American Legion in New Braunfels, Texas." Copy at Sophienburg Archives, New Braunfels, Texas, n.d.

"Looking at the Texas Highway System from all angles." Pamphlet, San Antonio: Texas Good Roads Association, ca. 1938, Center for American History, University of Texas at Austin.

"New Braunfels, 'The Beauty Spot of Texas,' Centennial Celebration." Official Souvenir Program, May 10-12, 1946, Center for American History, University of Texas at Austin.

"Night Scenes of Texas Centennial Exposition, Dallas, 1836-1936." Postcard Booklet, Dallas: Dallas Post Card Co., Made in C. T. Art-Colortone by Curt Teich & Co., Chicago, 1936, in author's possession.

"Plan For Highway Development." Booklet, New Braunfels, Texas, 1958, Center for American History, University of Texas at Austin.

"Pocket Facts." Pamphlet, Austin: Texas Department of Transportation, Winter 1997, March 2000.

Sanders, J. J. "Key to Comal County Road Marking American Legion Community Service Comal Post 179." Compiled by J. J. Sanders for Comal County, 1946. Copy located at Comal County Road Department, New Braunfels, Texas.

_____. Legion Scrapbook in author's possession.

_____. "Places Visited." Travel Diary in author's possession.

Sanders, Laurie P. Diary, 1946.

_____. Personal Scrapbook in author's possession.

"Texas and the Interstate Highway System." Booklet, Austin: Texas Highway Department, 1957, Center for American History, University of Texas at Austin.

"Texas Good Roads Association." Pamphlet, n.p.: Texas Good Roads Association, n.d., Center for American History, University of Texas at Austin.

Vertical Files. "Roads," File includes "State Department of Highways and Public Transportation, Introduction," Booklet, ca. 1986, Center for American History, University of Texas at Austin.

_____. "Texas Highway Department," Center for American History, University of Texas at Austin.

_____. "Travel Scrapbook," Center for American History, University of Texas at Austin.

"Waco Springs Park." Brochure, ca. early 1930s, Center for American History, University of Texas at Austin.

GOVERNMENT AND ORGANIZATIONAL RECORDS

American Legion Minutes. Comal Post 179, New Braunfels, Texas, September 30, 1929; January 14, 1931; February 5, 1931; 1932-33; July 10, 1932; August 9, 1932; March 20, 1933; May 26, 1933; January 6, 1936; April 6, 1936; August 3, 1939; June 10, 1940; July 8, 1940; April 14, 1941.

Court Minutes. Comal County Commissioners Court, Comal County Clerk's Office, New Braunfels, Texas, 1846-1980.

Inventory of County Archives. No. 46, Comal County, WPA Historical Records Survey, February 1937, Part 3, "Commissioners' Court," pp. 51-52. Copy at the Center for American History, University of Texas at Austin.

Road Minutes. Commissioners Court, Comal County, Texas, Comal County Clerk's Office, New Braunfels, Texas, 1882-1907, 1919.

United States War Department. Certificate for Joe Sanders, Government Request For Transportation, January 1919, in author's possession.

MASTERS THESES

Penshorn, Lillian E. "A History of Comal County." Southwest Texas State Teachers College, San Marcos, Texas, 1950.

Schumann, Iris Timmermann. "The Great Depression, 1929-1939: A Period of Changing Social Attitudes Toward Community Self-Reliance in New Braunfels and Comal County, Texas." Southwest Texas State University, San Marcos, Texas, 1980.

INDEX

The Author

✦ ✦ ✦

photo by Gary S. Hickinbotham

A native of New Braunfels, Texas, Laurie Jasinski graduated from Southwest Texas State University in 1987. She worked at the Texas State Historical Association in Austin on *The New Handbook of Texas* (1996) and has written features on Texana, tourism, and gardening for popular magazines such as *Texas Highways*, *Texas Parks & Wildlife*, and *Texas Gardener*. She has also contributed historical articles to *The Journal of South Texas*. Laurie and husband Gary S. Hickinbotham live in San Marcos, Texas.